Backyard Chickens

The Complete Beginner's Guide to Raising Healthy Hens for Fresh Eggs and Self-Sufficiency

Hudson Briggs

The Homestead Starter Series, Book 3

The Homestead Starter Series, Book 3

Disclaimer: This book is for general educational purposes only. Prices, regulations, growing conditions, animal ordinances, water rules, and food safety guidance vary by location and can change over time. Always confirm local laws and ordinances with the relevant authorities, and always follow current tested recipes from sources such as the USDA and the National Center for Home Food Preservation. This book is not a substitute for professional legal, financial, medical, veterinary, construction, or food-safety advice.

Table of Contents

A Note Before You Begin

This is Book 3 of The Homestead Starter Series. Book 1, The Beginner's Guide to Homesteading, introduced backyard chickens as one part of the bigger picture. This book is the complete guide: everything a beginner needs to bring home healthy birds, house them safely, feed them correctly, collect fresh eggs, and keep the flock thriving through every season and every problem.

Two things to know up front. First, this book is about keeping chickens, not about the carpentry of building a coop. You will learn exactly what a coop and run must provide and why, but the step-by-step build, with materials and cut lists, lives in Book 6, Homestead Projects. Second, the husbandry facts here, brooder temperatures, space requirements, feed protein levels, and the rest, are stated as the well-established figures used by experienced keepers and poultry extension sources, with ranges given where birds and conditions vary.

The usual disclaimer, said once: local laws, climate, and prices vary and change, so confirm your own rules and treat figures as realistic, checkable starting points. For any sick bird beyond basic care, a veterinarian or your local extension service is the right call.

Let's raise some chickens.

Introduction

What You'll Learn

This introduction explains why backyard chickens are one of the most rewarding steps a beginner can take, what keeping them really involves day to day, and how this book is organized to take you from an empty yard to a thriving, egg-producing flock.

Why Backyard Chickens

Chickens are the gateway animal of homesteading, and for good reason. They are small, inexpensive to keep, legal in more places than any other livestock, and they produce something you use every single day: fresh eggs. A small flock turns kitchen scraps, garden waste, and a modest amount of feed into better eggs than money can buy, along with rich manure for your garden and hours of genuine entertainment.

For a beginner, chickens hit a sweet spot. They are forgiving enough to learn on, productive enough to feel worthwhile almost immediately, and simple enough that the daily care takes only minutes. The first egg from your own hen is a small moment that makes the whole effort real.

What a Backyard Flock Provides - Eggs, Manure, Pest Control, Enjoyment

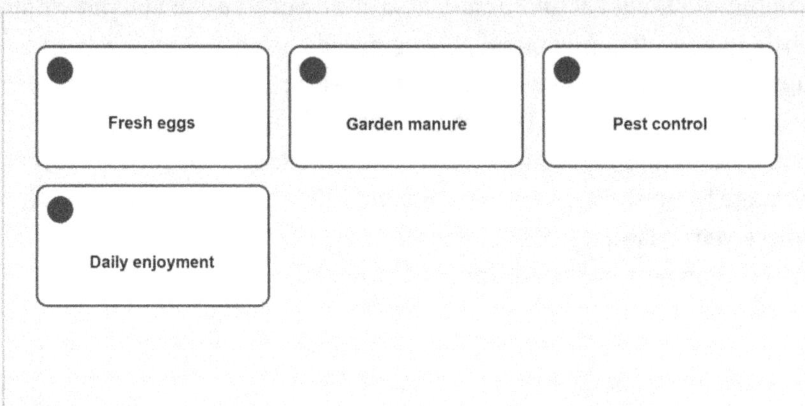

Why It Matters

Eggs are a staple most households buy every week, which makes them one of the few foods you can produce at home that replaces a constant, recurring purchase. Beyond the eggs, chickens connect you to your food, reduce your waste by eating scraps, fertilize your garden, and build the animal-keeping skills that make every other livestock project easier. They are the most practical first animal on almost any homestead.

What Keeping Chickens Really Involves

Be clear-eyed about the commitment before you start. Chickens need fresh food and water every day, a secure coop locked against predators every night, and clean bedding and a health check on a regular basis. They live for years, often well beyond their peak laying, so they are a multi-year responsibility, not a seasonal project. They also tie you down somewhat, since someone must care for them daily, including while you travel.

None of this is heavy work; the daily routine takes only a few minutes once set up. But it is daily, and going in with honest expectations is what separates a happy keeper from an overwhelmed one.

How This Book Works

The chapters follow the natural order of starting a flock. First you will handle the legal and practical groundwork and decide how many birds to keep. Then you will choose breeds, set up a safe coop and run, and bring home your chickens, whether as chicks or grown hens. From there you will learn to raise chicks, feed the flock correctly, understand and maximize egg production, keep your birds healthy, manage them through the seasons, handle flock behavior, and follow a first-year roadmap. Each chapter ends with a checklist and an action plan so you always know your next step.

Action Plan

Before the next chapter, do one thing: find out, with a single phone call or web search, whether your city or county allows backyard hens and how many. That one fact determines everything that follows, and it costs nothing but a few minutes.

Chapter 1: Before You Begin

What You'll Learn

This chapter covers the groundwork to settle before you buy a single bird: confirming what is legal where you live, understanding the real costs and commitment, and deciding how many hens to keep. Getting these right prevents the most common and most painful beginner mistakes.

Why It Matters

More backyard flocks are lost to skipped groundwork than to predators or disease. People buy adorable chicks on impulse, then discover their city bans hens, their lot is too small, or they cannot afford or manage the flock they bought. An hour of planning here saves money, heartbreak, and the grim task of rehoming birds you have grown attached to.

Confirm What's Legal

Before anything else, find out what your local rules allow. Three things matter most, and you can usually confirm all of them with one call to your city or county and a look at any homeowners association rules.

How many hens you may keep, which is often limited to somewhere between three and six in towns and suburbs. Whether roosters are allowed, which they usually are not in residential areas because of the noise. And whether there are setback rules dictating how far a coop must sit from your home and your neighbors' property lines, plus any permit requirement for the coop itself.

Write down the answers and who gave them to you. If you are in a homeowners association, read its rules too, because they can be stricter than city law and they win on their own turf.

Three Legal Questions to Answer Before Getting Hens

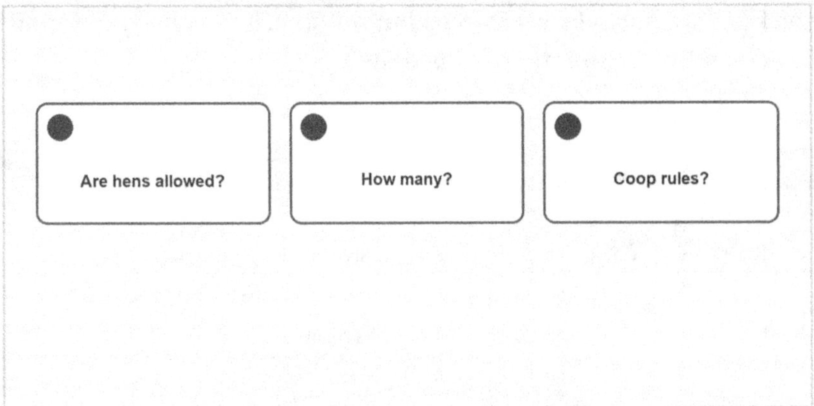

Understand the Commitment

Chickens are a years-long, daily responsibility. A hen commonly lives six to ten years or more, while laying strongly for only the first two to three, which means you will likely keep and feed birds well past their most productive years. Plan for that from the start, and decide in advance how you will handle older, non-laying hens, whether you will keep them as pets, which is perfectly valid, or manage the flock another way.

Day to day, someone must feed, water, and lock up the flock every day without exception, including weekends, holidays, and trips. Line up a reliable chicken-sitter before you travel. The work is light, but it does not take days off.

Understand the Costs

Chicken-keeping has two kinds of cost: the one-time setup and the ongoing care. Setup is the larger number and includes the coop and run, feeders and waterers, and the birds themselves. Ongoing cost is mainly feed, plus bedding and occasional health supplies.

The single biggest variable is the coop. Build one yourself from reclaimed materials and it can be nearly free; buy a large ready-made one and it can cost many hundreds of dollars. Feed is a steady, predictable cost covered in detail in Chapter 7. The honest summary: chickens are inexpensive to keep but not free, and the eggs pay you back over time rather than immediately.

Table: Where the Money Goes

Cost	Type	Notes
Coop and run	One-time	Biggest variable; cheap if built, costly if bought
Feeders and waterers	One-time	Inexpensive and long-lasting
The birds	One-time	Chicks are cheapest, grown hens cost more
Feed	Ongoing	The main recurring cost
Bedding	Ongoing	Modest; straw, shavings, or similar
Health supplies	Occasiona l	Grit, oyster shell, basic first aid items

Decide How Many Hens

Three questions set your flock size: how many eggs your household wants, how much space and budget you have, and what your local limit allows.

A useful rule of thumb is about two hens per person in the household for a steady egg supply, since a good layer produces roughly four to six eggs a week in her productive months. For a family of four, that points to roughly six to eight hens, eggs to spare included. Always keep at least three birds, never just one or two, because chickens are flock animals that become stressed and unhappy alone.

Then check that number against your space, using the requirements in Chapter 3, and against your legal limit. The smaller of what you want, what fits, and what is allowed is your answer.

Matching Flock Size to Household and Space

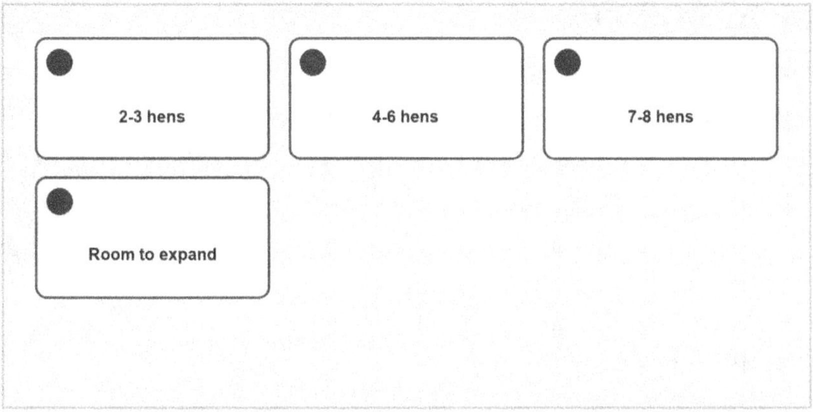

When in doubt, start at the lower end; it is far easier to add a few birds next year than to manage too many in too little space.

Table: How Many Hens to Start With

Household	Eggs Wanted	Suggested Hens
1 to 2 people	A few eggs most days	3 to 4 hens
3 to 4 people	About a dozen a week, with extra	6 to 8 hens
5+ people	A dozen or more a week	8 or more, within local limits

Common Mistakes

Buying chicks before confirming they are legal where you live, then having to rehome them.

Ignoring the years-long commitment and being unprepared for hens that outlive their best laying years.

Underestimating setup cost, especially the coop, and running out of money mid-project.

Keeping only one or two birds, which leaves social flock animals stressed and lonely.

Starting with far too many birds for the available space, causing crowding, stress, and disease.

Having no plan for daily care during travel.

Budget Options

The free or near-free path: Confirm your rules at no cost, plan to build a coop from reclaimed materials, and start with the minimum healthy flock of three to four hens to keep both setup and feed costs low.

The starter path: A modest ready-made or self-built coop, basic equipment, and a small starter flock is a reasonable one-time investment that fits most beginner budgets and pays back in eggs over the following years.

Quick Wins

Confirm your legal hen limit and rooster rule today.

Read your homeowners association rules if you have one.

Decide your flock size using the two-hens-per-person guideline and your local limit.

Line up who will care for the flock when you travel.

Chapter Checklist

[ ] I have confirmed my legal hen limit, rooster rule, and setbacks

[ ] I have read my HOA rules if applicable

[ ] I understand chickens are a years-long daily commitment

[ ] I have a realistic picture of setup and ongoing costs

[ ] I have chosen a flock size that fits my eggs, space, and legal limit

[ ] I have at least three birds planned, never one or two

[ ] I have a plan for care during travel

Action Plan

This week, settle the groundwork. Confirm exactly how many hens you may keep and whether roosters are allowed. Decide your flock size using the two-hens-per-person guideline, capped by your legal limit and your space. Sketch a rough budget for setup and feed, and line up a chicken-sitter for travel. With the foundation set, the next chapter helps you choose the right breeds for your eggs, your climate, and your temperament.

Chapter 2: Choosing the Right Breeds

What You'll Learn

This chapter helps you choose breeds that fit your goals, your climate, and your temperament. You will learn the difference between egg-laying, dual-purpose, and ornamental breeds, which traits actually matter for a beginner, and which specific breeds are the most reliable, friendly, and productive first chickens.

Why It Matters

The breeds you choose shape your entire experience. The right ones lay well, tolerate your weather, and are calm and easy to handle. The wrong ones may be flighty, fragile in your climate, or poor layers, turning an easy hobby into a frustrating one. Choosing well at the start avoids years of mismatch.

What to Look For in a Breed

Four traits matter most for a beginner. Egg production tells you how many eggs to expect; the best layers give 250 or more a year, while ornamental breeds may give only a hundred. Temperament matters enormously for a backyard flock handled by a family; you want calm, docile, friendly birds, not nervous or aggressive ones. Climate hardiness determines whether a breed thrives where you live; heavy, well-feathered breeds handle cold, while lighter breeds tolerate heat. And purpose, whether you want eggs only, meat too, or simply pets, narrows the field.

A beginner is almost always best served by calm, hardy, productive, dual-purpose breeds or by reliable laying hybrids.

Four Traits to Weigh When Choosing a Breed

- Temperament
- Egg production
- Climate fit
- Hardiness

Layers, Dual-Purpose, and Ornamental

Breeds fall roughly into three groups. Laying breeds are bred for maximum eggs and are usually lighter, more active birds; the white-egg Leghorn is the classic example. Dual-purpose breeds are sturdy, calm birds that lay well and also have enough size to be useful for meat; most of the best beginner breeds fall here. Ornamental and bantam breeds are kept mainly for looks, personality, or as pets, and they lay fewer eggs; the fluffy Silkie is a favorite example, prized as a gentle pet and a determined mother hen rather than a producer.

The Best Beginner Breeds

The breeds below are widely recommended for beginners because they combine reliable laying, calm temperament, and hardiness. Egg numbers are realistic annual ranges; real output depends on the individual bird, her age, the season, and her care.

Table: Reliable, Friendly Breeds for First-Time Keepers

Breed	Eggs per Year	Egg Color	Temperament	Climate Notes
Rhode Island Red	200 to 280	Brown	Hardy and friendly	Cold-hardy, very adaptable
Australorp	250 to 300	Brown	Gentle and quiet	Cold-hardy, tolerates heat reasonably
Plymouth Barred Rock	200 to 280	Brown	Calm and curious	Very cold-hardy
Buff Orpington	180 to 240	Brown	Very docile, can go broody	Cold-hardy, heavy feathering
Sex-link hybrid	250 to 320	Brown	Calm and easy	Adaptable; shorter productive life
Leghorn	250 to 300	White	Active, can be flighty	Heat-tolerant, top white-egg layer
Easter Egger	200 to 280	Blue or green	Friendly and curious	Cold-hardy, fun colored eggs
Wyandotte	180 to 260	Brown	Calm and dignified	Excellent cold tolerance
Silkie (bantam)	100 to 120	Cream or tinted	Extremely docile, broody	Pet and mothering breed, poor layer

A note on choosing from this table: for the most eggs with the least fuss, the Australorp, sex-link hybrids, and Rhode Island Red are hard to beat. For colored eggs and a family-friendly personality, Easter Eggers are a delight.

Egg Color Variety by Breed

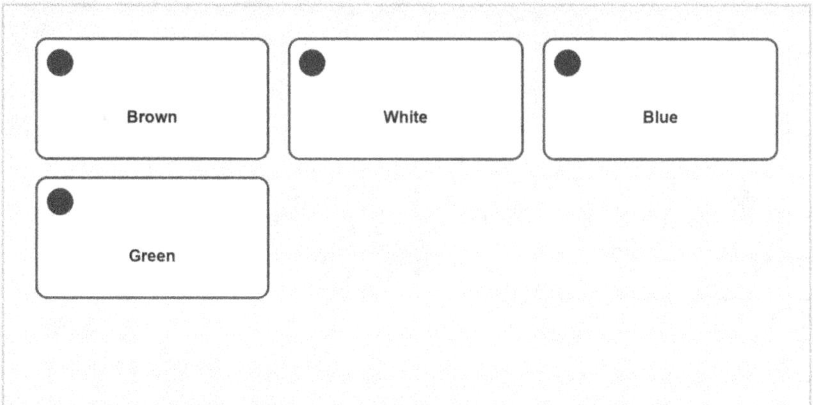

For the coldest climates, the Wyandotte and Barred Rock excel. For a gentle pet or a natural mother hen, the Silkie is unmatched, just do not expect many eggs.

Mixing Breeds

You can absolutely keep a mixed flock, and many beginners enjoy the variety of different colors, personalities, and egg shades. The main guideline is to mix breeds of similar size and temperament, so a flock of calm, standard-sized hens lives in harmony, while mixing very small, timid birds with large, assertive ones can lead to bullying. A mix of several of the calm beginner breeds above makes a colorful, peaceful, productive flock.

Match the Breed to Your Climate

Climate is the trait beginners most often overlook. In cold regions, choose heavy, well-feathered, cold-hardy breeds and be aware that large single combs can be prone to frostbite in extreme cold; smaller-combed breeds fare better. In hot regions, lighter, more active breeds like the Leghorn shed heat better, and all birds will need shade and ample water. Your local feed store tends to stock breeds suited to your area, which is a quiet advantage of buying locally.

Matching Breeds to Cold and Hot Climates

- Cold hardy
- Heat tolerant
- Comb type
- Shade needs

Common Mistakes

Choosing a breed on looks alone, then discovering it is flighty, fragile in your climate, or a poor layer.

Ignoring climate and keeping heat-loving breeds in harsh winters or heavy breeds in extreme heat.

Mixing very different sizes and temperaments, which invites bullying in the flock.

Expecting ornamental or bantam breeds to lay like production hens.

Keeping a rooster by accident because chicks were not sexed, in an area that bans them.

Budget Options

The free or near-free path: Choose hardy, dual-purpose breeds that thrive on less and forage well, reducing feed needs, and buy locally stocked breeds suited to your climate to avoid losses.

The starter path: A mix of a few proven beginner breeds, chosen for laying and temperament, gives a productive, enjoyable flock and is worth selecting deliberately rather than grabbing whatever is available.

Quick Wins

List your priorities: eggs, temperament, climate hardiness, or colored eggs.

Pick two or three calm, productive breeds suited to your climate from the table.

Check which breeds your local feed store stocks for your area.

If you want hens only, plan to buy sexed pullets rather than unsexed chicks.

Chapter Checklist

[ ] I know which traits matter most for my situation

[ ] I understand the difference between laying, dual-purpose, and ornamental breeds

[ ] I have chosen breeds suited to my climate

[ ] My chosen breeds are calm and beginner-friendly

[ ] If mixing breeds, I have matched size and temperament

[ ] I have a plan to get hens, not accidental roosters

Action Plan

Decide your top priorities, eggs, temperament, climate, or egg color, and pick two or three breeds from the comparison table that fit. Check what your local feed store carries, since locally stocked breeds suit your climate and reduce losses. With your breeds chosen, the next step is giving them a safe, comfortable home, which begins with the coop.

Chapter 3: The Coop

What You'll Learn

This chapter covers what a good coop must provide: enough space, proper nest boxes and roosts, the ventilation that keeps birds healthy, suitable bedding, and a secure, cleanable design. You will learn the requirements and the reasoning so you can build or buy the right coop. The actual construction, with materials and cut lists, is in Book 6, Homestead Projects.

Why It Matters

The coop is your flock's bedroom and shelter, where they sleep, lay, and take refuge from weather and predators. A good coop keeps birds healthy, safe, and laying; a poor one causes stress, disease, frostbite, and predator losses. Getting the requirements right, especially space and ventilation, prevents most of the problems beginners face.

How Much Space

Crowding is one of the most common and damaging beginner mistakes, causing stress, feather-picking, bullying, and disease. The standard guideline is about 4 square feet of coop floor space per standard-sized hen, with bantams needing a bit less. This is the indoor sleeping and laying space; the outdoor run is separate and covered in the next chapter.

More space is always better than less. If you are deciding between two coop sizes, choose the larger, because a flock almost always grows and crowding causes far more problems than extra room ever does.

Inside a Well-Designed Coop

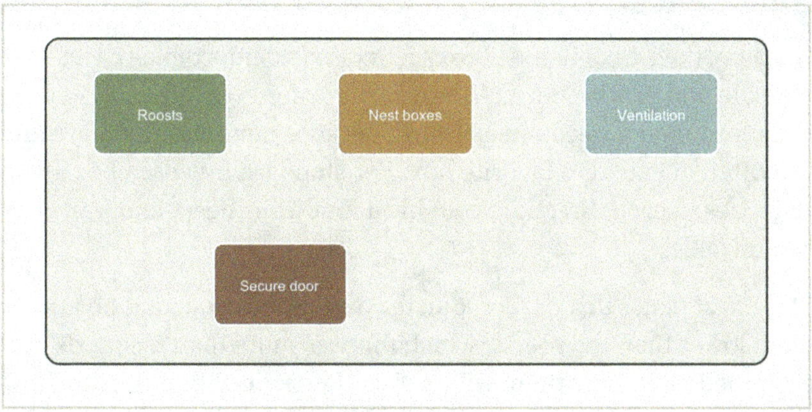

Table: Minimum Coop Floor Space

Hens	Minimum Coop Space	Notes
3 hens	12 square feet	At 4 sq ft per standard hen
4 hens	16 square feet	More is better
6 hens	24 square feet	Plan for flock growth
8 hens	32 square feet	Generous space prevents problems

Coop Floor Space Requirements by Flock Size

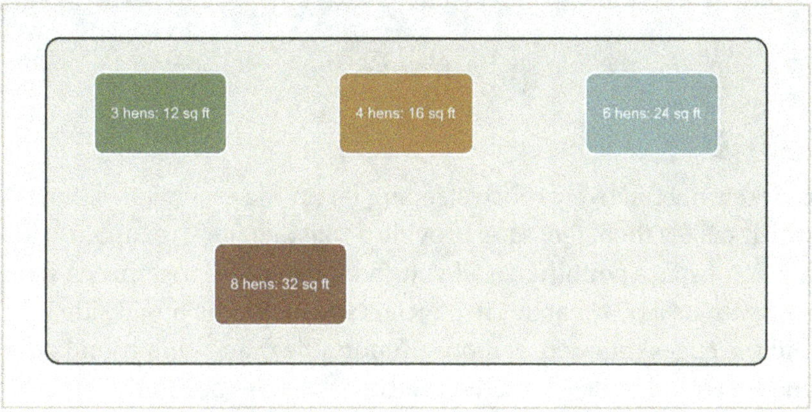

Nest Boxes

Hens lay their eggs in nest boxes, enclosed, comfortable spaces that feel safe and private. Provide one nest box for every three to four hens; you do not need one per bird, because hens happily share and will often queue for a favorite box. For standard breeds, a box about 12 inches in each direction, roughly a foot wide, deep, and tall, is comfortable.

Place nest boxes in a darker, quieter part of the coop, and position them lower than the roosts, which matters more than it sounds, as explained next. Line them with soft bedding, keep that bedding clean, and collect eggs daily so they stay clean and unbroken.

Nest Box Size and Placement

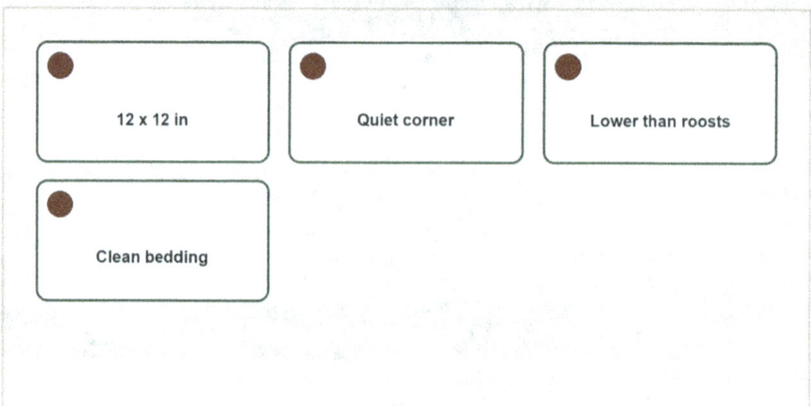

Roosts

Chickens instinctively roost, sleeping up off the ground on a bar, which makes them feel safe. Provide a roosting bar with about 8 to 12 inches of space per bird, so all your hens can perch together. A flat bar, such as a board about two inches wide with the broad side up and the edges rounded, is more comfortable than a thin round pole and, in cold climates, lets a hen settle down over her feet to keep her toes warm and prevent frostbite.

The single most important roost rule: place roosts higher than your nest boxes. Chickens want to sleep at the highest point available, so if the nest boxes are higher than or level with the roosts, the hens will sleep in the boxes, soiling them and the eggs. Roosts up high, nest boxes lower, keeps the eggs clean.

Roosts Higher Than Nest Boxes

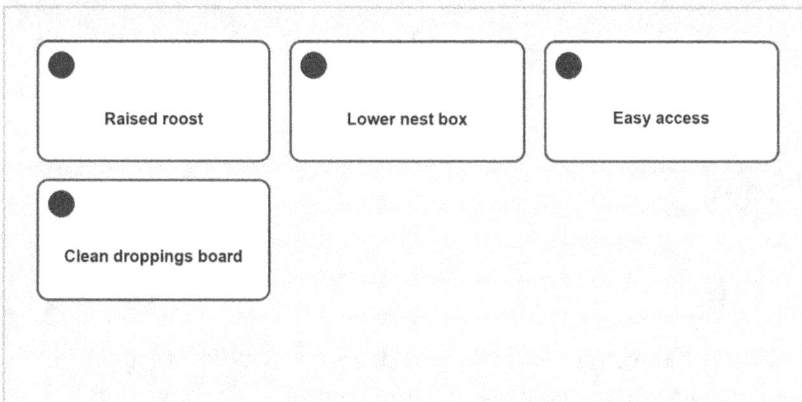

Ventilation

Ventilation is the most misunderstood and most important feature of a coop. Chickens give off a surprising amount of moisture through their breath and droppings, and that moisture, combined with ammonia from droppings, must escape or it causes respiratory illness and, in cold weather, frostbite. Damp air, not cold air, is the real enemy in winter.

The rule is good ventilation without drafts at roost level. Place vents up high, above where the birds roost, so stale, moist air rises and escapes while the birds themselves sit in still air. A well-ventilated coop stays dry and breathable even when closed against the cold. Sealing a coop up tight to keep it warm is a classic and harmful mistake; it traps moisture and makes birds sick.

Coop Ventilation Without Drafts

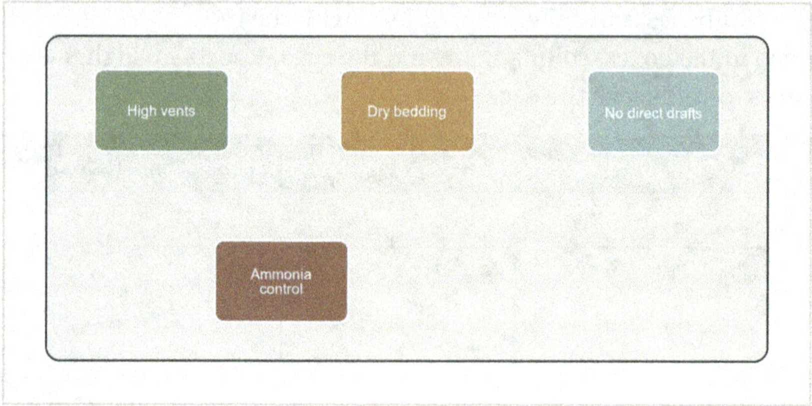

Bedding

Bedding on the coop floor absorbs droppings and moisture and keeps the coop dry and clean. Common choices include pine shavings, straw, or hemp bedding. Keep it dry and refresh or replace it before it becomes damp or smelly. Many keepers use the deep litter method, adding fresh bedding on top of old and letting the lower layers slowly compost in place over a season, which reduces labor and produces excellent garden compost, then doing a full clean-out periodically. Whatever method you use, a dry coop is a healthy coop.

Security and Cleanability

A coop must be genuinely predator-proof, because predators are persistent and clever, especially at night. This means solid construction with no gaps a raccoon can reach through or a weasel can squeeze into, secure latches that a raccoon's dexterous paws cannot open, and hardware cloth rather than flimsy wire over any openings. Predator protection is covered fully in the next chapter, but the coop is the last line of defense and must be solid.

Design for easy cleaning, too. A coop you can access easily, with a removable droppings tray or a door you can muck out through, is a coop you will actually keep clean, and clean is healthy. These practical features are exactly what the build plans in Book 6 are designed around.

Common Mistakes

Building too small and crowding the flock, which causes stress, picking, and disease. Err larger.

Sealing the coop tight against cold and trapping moisture, causing respiratory illness and frostbite. Ventilate up high.

Placing nest boxes higher than roosts, so hens sleep in and soil the boxes.

Using flimsy wire or weak latches that predators defeat. Use hardware cloth and secure latches.

Making the coop hard to clean, so it gets neglected and the flock gets sick.

Letting bedding get damp, which breeds illness. Keep the coop dry.

Budget Options

The free or near-free path: Build a coop from reclaimed and salvaged materials, which can cost almost nothing while still meeting every requirement for space, ventilation, and security. The plans in Book 6 are designed to be built affordably.

The starter path: A quality ready-made coop sized generously for your flock saves effort, but check that it truly provides the space and ventilation claimed, since many commercial coops are rated for far more birds than they should hold.

Quick Wins

Calculate your minimum coop size at 4 square feet per hen, then plan bigger.

Plan one nest box per three to four hens, placed low and dark.

Plan roosts higher than the nest boxes, with rounded flat bars.

Plan high vents for ventilation without roost-level drafts.

Chapter Checklist

[ ] My coop provides at least 4 square feet per standard hen, ideally more

[ ] I have one nest box per three to four hens, placed low

[ ] My roosts are higher than the nest boxes, with comfortable flat bars

[ ] My coop has high ventilation without drafts at roost level

[ ] I have a plan for dry, clean bedding

[ ] My coop is solidly predator-proof and easy to clean

Action Plan

Work out your coop requirements on paper: total floor space at 4 square feet per hen with room to spare, the number and placement of nest boxes, roost height and length, and high ventilation. Decide whether to build or buy, checking any commercial coop honestly against these numbers. When you are ready to construct one, Book 6 provides a beginner-proof build. With the coop planned, the next chapter secures the run and defends the flock from predators.

Chapter 4: The Run and Predator Protection

What You'll Learn

This chapter covers the outdoor run where your hens spend their days and, just as importantly, how to protect them from the predators that will otherwise find them. You will learn how much run space hens need, how to predator-proof properly, the trade-offs of free-ranging, and which predators to guard against.

Why It Matters

This is where beginners lose birds. Almost every predator in your area, and several you would not expect, will happily eat a chicken given the chance, and a single breach can wipe out an entire flock in one night. Predator protection is not optional or something to add later; it is built in from day one. A secure run is the difference between a thriving flock and a heartbreaking loss.

The Run: Outdoor Space

The run is the enclosed outdoor area attached to the coop, where hens scratch, dust-bathe, forage, and get fresh air and exercise. Provide about 8 to 10 square feet of run space per standard hen, and as with the coop, more is better. Cramped runs lead to bare, muddy ground, boredom, and the behavior problems that come with it, while a roomy run keeps birds active, healthy, and content.

Run Space Requirements by Flock Size

- 3 hens: 24-30 sq ft
- 4 hens: 32-40 sq ft
- 6 hens: 48-60 sq ft
- 8 hens: 64-80 sq ft

Table: Minimum Run Space

Hens	Minimum Run Space	Notes
3 hens	24 to 30 square feet	At 8 to 10 sq ft per hen
4 hens	32 to 40 square feet	More space means healthier birds
6 hens	48 to 60 square feet	Prevents bare, muddy ground
8 hens	64 to 80 square feet	Roomy runs reduce behavior problems

Predator-Proofing: The Three Rules

Three rules prevent the large majority of predator losses, and they are non-negotiable.

First, use hardware cloth, never chicken wire. This is the most important single fact in this chapter. Chicken wire is designed to keep chickens in, not predators out; a determined raccoon can tear it or reach through it, and small predators slip through its gaps. Half-inch hardware cloth is the standard for covering all openings and enclosing the run.

Hardware Cloth vs Chicken Wire

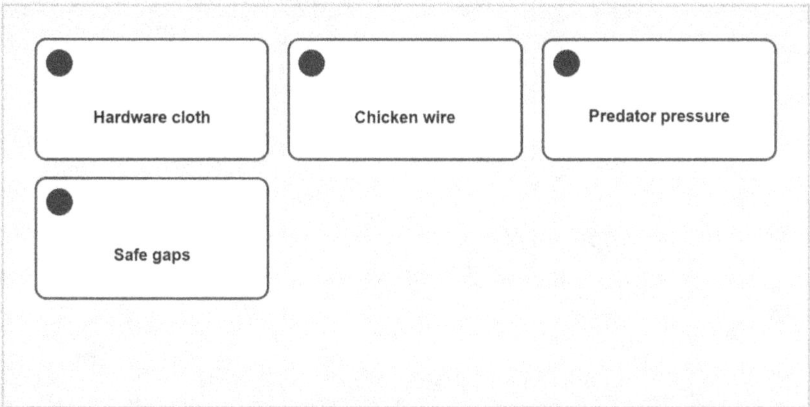

Second, block digging predators with a buried apron. Foxes, dogs, and others dig under fences. Extend hardware cloth downward and outward at the base of the run, either buried about a foot deep or laid flat outward along the ground as a skirt about a foot or more wide, so a digging animal hits wire instead of soft earth.

Third, cover the run and lock up at night. Hawks and owls attack from above and raccoons climb, so an open-topped run is vulnerable; cover it with wire or netting. And because most predators hunt after dark, lock your flock securely in the coop every single night. An automatic coop door is worth every penny for the nights you forget or come home late.

Predator-Proofing - Hardware Cloth, Buried Apron, Covered Run

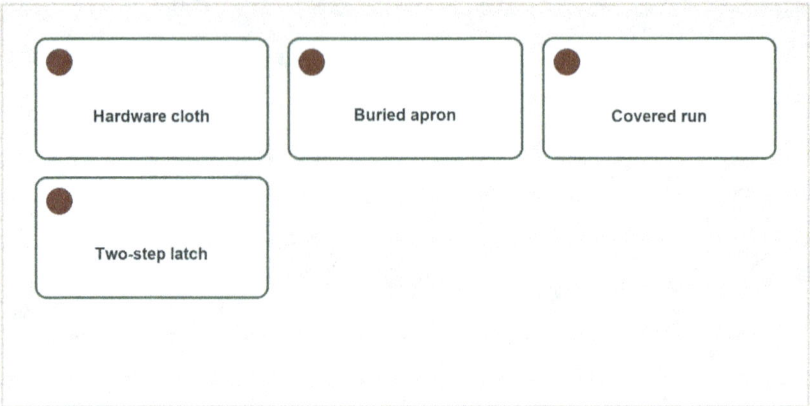

Know Your Predators

Different predators attack differently, and knowing the threats in your area helps you defend against them. The table below covers the common ones.

Table: Threats to Guard Against

Predator	Threat	Key Defense
Raccoon	Reaches through wire, opens latches, climbs	Hardware cloth, secure latches, lock at night
Fox and coyote	Digs and grabs, mostly at dawn and dusk	Buried apron, secure run, locked coop
Hawk and owl	Attacks from above	Covered run, overhead netting
Dog	Digs, breaks in, kills many at once	Strong fencing, buried apron
Weasel and mink	Squeezes through tiny gaps	Half-inch hardware cloth, no gaps
Snake	Eats eggs and chicks	Tight hardware cloth, collect eggs
Opossum and rat	Eats eggs, chicks, and feed	Secure coop, store feed in metal bins

A note on latches: raccoons have remarkably dexterous paws and can open simple hooks and slide bolts. Use latches that require two steps or a twist, the kind a raccoon cannot work, on every coop door.

Free-Ranging: The Trade-Offs

Letting hens roam your yard freely, called free-ranging, has real benefits: happier, more active birds, lower feed costs as they forage for bugs and greens, and natural pest control for your garden. But it carries real risks, too: far greater predator exposure during the day, including from hawks, and hens that range into gardens, neighbors' yards, or roads.

A popular middle path is supervised or partial free-ranging, letting birds out into a yard or larger fenced area only when you are around or for a few hours before dusk, then securing them in the run and coop otherwise. This captures much of the benefit with less of the risk. Whatever you choose, the coop must always be locked at night, when predators are most active.

Free-Ranging Benefits and Risks

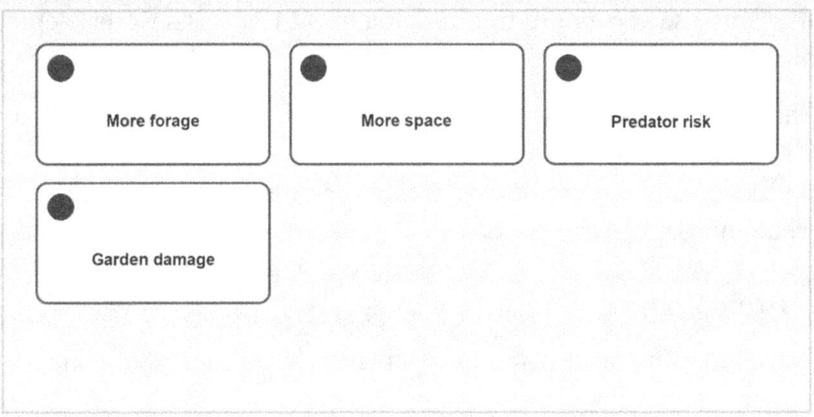

Common Mistakes

Using chicken wire instead of hardware cloth, the single most common fatal mistake.

Leaving the run open on top, exposing birds to hawks, owls, and climbing predators.

Failing to block digging with a buried apron, letting foxes and dogs tunnel in.

Using simple latches a raccoon can open. Use two-step or twist latches.

Forgetting to lock up at night even once, which is often all a predator needs.

Free-ranging unsupervised in high-predator areas and suffering steady losses.

Storing feed where it attracts rats and other pests. Use sealed metal containers.

Budget Options

The free or near-free path: Build the run from salvaged framing covered in hardware cloth, which is the one component worth buying new, and use supervised free-ranging to cut feed costs while keeping birds safe.

The starter path: A well-built covered run in hardware cloth with a buried apron and an automatic coop door is the gold standard for protection and pays for itself the first time it stops a predator that would have taken the flock.

Quick Wins

Calculate your run size at 8 to 10 square feet per hen, then plan bigger.

Commit to hardware cloth everywhere, never chicken wire.

Plan a buried apron and a covered top for the run.

Install two-step or twist latches a raccoon cannot open.

Chapter Checklist

[ ] My run provides at least 8 to 10 square feet per hen

[ ] I am using half-inch hardware cloth, not chicken wire

[ ] I have a buried apron or skirt against digging predators

[ ] My run is covered against aerial and climbing predators

[ ] My coop has raccoon-proof latches and is locked every night

[ ] I store feed in sealed containers away from pests

[ ] I have a sensible free-ranging plan for my predator risk

Action Plan

Plan your run at 8 to 10 square feet per hen, fully enclosed in half-inch hardware cloth with a buried apron and a covered top. Fit raccoon-proof latches and, if you can, an automatic coop door so the flock is locked up reliably every night. Decide your free-ranging approach based on your predator risk. With a safe coop and run ready, it is finally time to bring home your chickens, which the next chapter covers.

Chapter 5: Getting Your Chickens

What You'll Learn

This chapter helps you decide how to start your flock, by chicks, by grown hens, or by hatching your own, and where to get healthy birds. You will learn the trade-offs of each starting point, what a healthy bird looks like, and the key terms like sexed and straight run that you need to understand before you buy.

Why It Matters

How you start shapes your first months. Chicks are cheap and bond with you but take months to lay and need special care. Grown hens lay right away but cost more and come with unknown history. Choosing the right starting point for your situation, and getting healthy stock from a good source, sets your flock up to thrive and avoids importing disease or accidental roosters.

Three Ways to Start

You have three practical options for beginning a flock.

Day-old chicks are the most common and least expensive way to start, and raising them is rewarding and builds a bond. The trade-off is that they need a brooder with careful heat and attention for several weeks, covered in the next chapter, and they will not lay until they mature, at roughly five to six months. This is the path most beginners take and enjoy.

Started or point-of-lay pullets are young female birds already grown past the fragile chick stage, sometimes right at the age of first laying. They cost more than chicks but skip the brooder stage and begin laying much sooner. They are an excellent choice if you want eggs without raising chicks.

Grown hens give you eggs immediately, but they cost the most per bird, may be past their laying prime, and come with a history you cannot fully verify, so health and disease risk matter. Hatching your own eggs in an incubator is a fourth route, but it is more advanced, yields roughly half roosters, and is best left until you have experience.

Table: Comparing Your Starting Options

Starting Point	Cost	Time to Eggs	Best For
Day-old chicks	Lowest	About 5 to 6 months	Most beginners, bonding, lowest cost
Started or point-of-lay pullets	Moderate	Weeks or immediate	Eggs sooner without the brooder stage
Grown laying hens	Highest per bird	Immediate	Instant eggs, accept unknown history
Hatching eggs in an incubator	Low per egg, more gear	About 6 months	Experienced keepers; expect many roosters

Four Ways to Start a Flock Compared

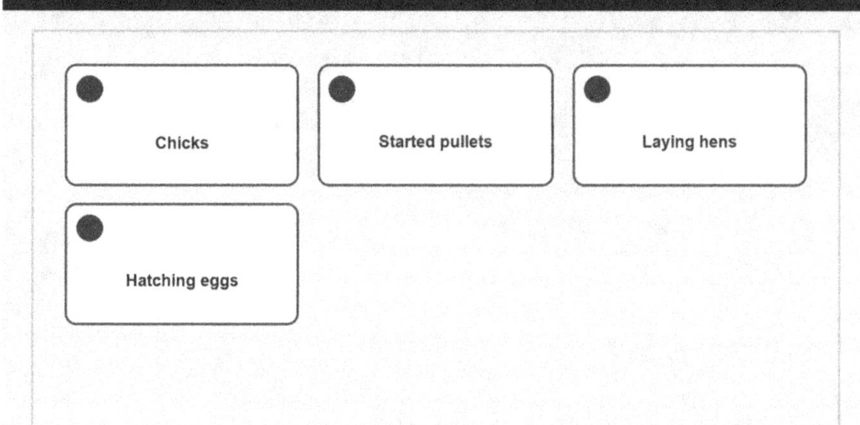

Sexed or Straight Run

When buying chicks, you will see two terms that matter enormously. Sexed chicks, sold as pullets, have been identified as female, so you get hens, which is what you want if roosters are banned or unwanted. The sexing is highly accurate but not perfect, so an occasional rooster slips through. Straight run chicks are unsexed, sold as they hatch, which means roughly half will turn out to be roosters. Straight run is cheaper, but for a beginner in a typical backyard, especially where roosters are illegal, buy sexed pullets and avoid the problem.

Sexed Pullets vs Straight Run

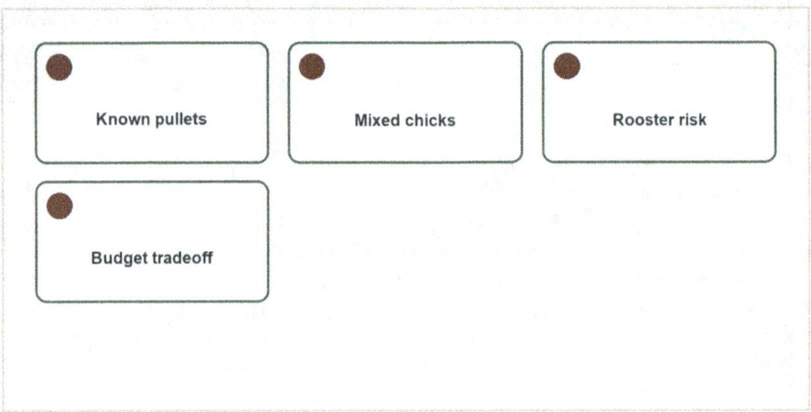

Where to Get Healthy Birds

Source matters, both for the health of your birds and to avoid bringing disease into your yard. Reputable feed stores in spring stock chicks from hatcheries and are a convenient, reliable starting point. Established hatcheries ship healthy, often vaccinated chicks and offer wide breed selection, usually with a minimum order. Reputable local breeders can be excellent for specific breeds. What to avoid is buying birds of unknown health from informal sources like swap meets, where disease spreads easily.

Wherever you buy, you can often request that chicks be vaccinated for Marek's disease, a common and serious poultry illness, which is done at the hatchery in the first day of life. This is worth doing and covered more in the health chapter.

What a Healthy Bird Looks Like

Whether choosing chicks or grown birds, look for signs of health. A healthy bird is active, alert, and curious, not lethargic or huddled. Its eyes are bright and clear, not cloudy or weepy, and its nostrils are clean with no discharge. Its feathers, in grown birds, are clean and reasonably smooth, and the vent area at the rear is clean, not pasted or dirty. The bird stands and moves normally on good legs and feet. In chicks, avoid any that are listless, have a pasted-over vent, or seem weak. Choosing healthy stock from the start prevents a great deal of trouble.

Signs of a Healthy Chicken

- Bright eyes
- Clean vent
- Smooth feathers
- Active stance

Timing

Spring is the classic time to start, when feed stores stock chicks and the warming weather suits raising them, so they mature into laying by late summer or fall. You can start at other times with more attention to heat in a brooder, but for a first flock, spring is the easiest and most forgiving season.

Common Mistakes

Buying straight run chicks where roosters are banned, then ending up with illegal roosters to rehome.

Getting birds from informal, unverified sources and importing disease into the flock.

Choosing the cheapest birds without checking health, then dealing with sick stock.

Starting with grown hens of unknown history and inheriting hidden health or laying problems.

Buying chicks before the brooder is ready, leaving fragile birds without proper heat.

Budget Options

The free or near-free path: Day-old sexed chicks are the cheapest healthy way to start, especially bought locally in spring, and they reward you with a bond and the full experience of raising your flock.

The starter path: Started or point-of-lay pullets cost more but skip the brooder stage and bring eggs sooner, a worthwhile trade for those who want a faster, simpler start.

Quick Wins

Decide your starting point: chicks, started pullets, or grown hens.

If buying chicks, commit to sexed pullets unless roosters are allowed and wanted.

Identify a reputable source: a feed store, established hatchery, or known breeder.

Plan to start in spring if you can, and request Marek's vaccination.

Chapter Checklist

[ ] I have chosen how to start: chicks, started pullets, or grown hens

[ ] If buying chicks, I am getting sexed pullets unless I want roosters

[ ] I have a reputable source lined up

[ ] I know how to spot a healthy bird

[ ] I plan to start in spring if possible

[ ] I will request Marek's vaccination for chicks

Action Plan

Decide your starting point based on how soon you want eggs and whether you want to raise chicks. Line up a reputable source, a spring feed store, an established hatchery, or a known breeder, and plan for sexed pullets unless roosters are welcome. If you are starting with chicks, do not bring them home until the brooder is ready, which the next chapter shows you how to set up.

Chapter 6: Raising Chicks

What You'll Learn

This chapter walks you through raising day-old chicks into healthy young birds: setting up a safe brooder, providing the right heat on the right schedule, feeding and watering chicks correctly, watching for the common chick problems, and knowing when they are ready to move outside. Get the first few weeks right and the rest is easy.

Why It Matters

Chicks are fragile in their first weeks and depend entirely on you for warmth, clean food and water, and a safe space. The two things that matter most, correct heat and clean water, are simple to provide once you know the targets. Most chick losses come from getting heat wrong, from drowning in deep water, or from a preventable problem caught too late, all of which this chapter helps you avoid.

Setting Up the Brooder

The brooder is the warm, safe, enclosed space where chicks live for their first several weeks, before they are feathered enough to handle the outdoors. It needs four things: a heat source, a draft-free enclosure, bedding, and food and water.

The enclosure can be as simple as a large tote, a stock tank, or a sturdy cardboard or wood box, with walls high enough that chicks cannot hop out as they grow. Allow enough space, roughly half a square foot per chick at first, increasing as they grow, since crowded chicks pile up and overheat. Keep it in a draft-free, predator-safe spot such as a garage, spare room, or outbuilding.

For bedding, use pine shavings, which are absorbent and safe. Avoid cedar shavings, whose fumes can harm chicks, and avoid slippery surfaces like newspaper, which can cause leg problems. Keep bedding dry and clean.

A Basic Brooder Setup for Chicks

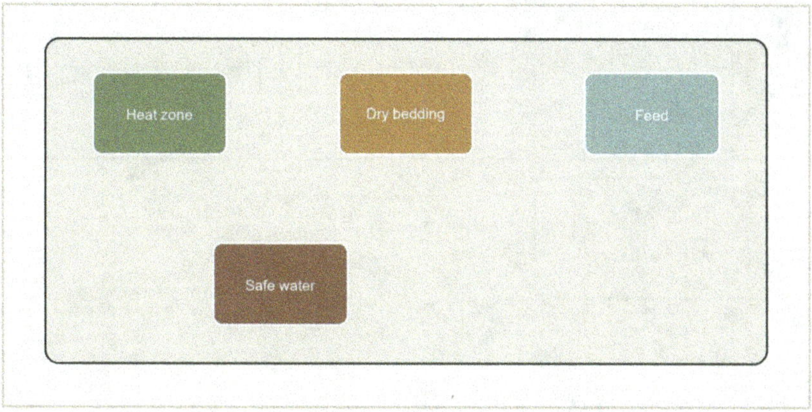

Heat: The Most Important Detail

Chicks cannot regulate their own temperature at first and need a warm zone to thrive. There are two ways to provide it. A heat lamp is traditional but carries a real fire risk, as the bulbs run extremely hot and have started many barn and house fires; if you use one, secure it in multiple ways so it cannot fall. A safer modern alternative is a brooder plate, a low, flat heater the chicks huddle under like they would under a mother hen, which uses less power and poses little fire risk. Brooder plates are strongly recommended for beginners.

The temperature schedule is straightforward. Keep the warm zone at about 90 to 95 degrees Fahrenheit for the first week, then reduce it by roughly 5 degrees each week, until at around five to six weeks the chicks are acclimated to ambient temperature, near 70 degrees. Always provide the heat in one zone, not the whole brooder, so chicks can move toward or away from it to find their comfort.

One Warm Zone in the Brooder

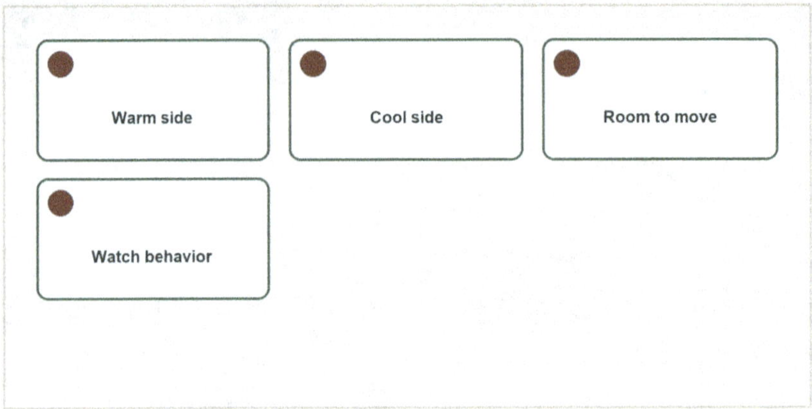

The chicks themselves tell you if the temperature is right. If they huddle tightly together directly under the heat, they are too cold. If they crowd to the far edges away from the heat and pant, they are too hot. If they are spread comfortably around the brooder, moving freely, the temperature is just right. Watch their behavior as your true guide, with the thermometer as a backup.

Table: Heat Targets by Age

Chick Age	Warm Zone Temperature	Notes
Week 1	90 to 95 degrees F	Watch chick behavior to fine-tune
Week 2	85 to 90 degrees F	Reduce about 5 degrees
Week 3	80 to 85 degrees F	Chicks feathering out
Week 4	75 to 80 degrees F	Continue reducing
Week 5 to 6	About 70 degrees F or ambient	Ready to wean off heat when feathered

Reading Chick Behavior - Too Hot, Too Cold, Just Right

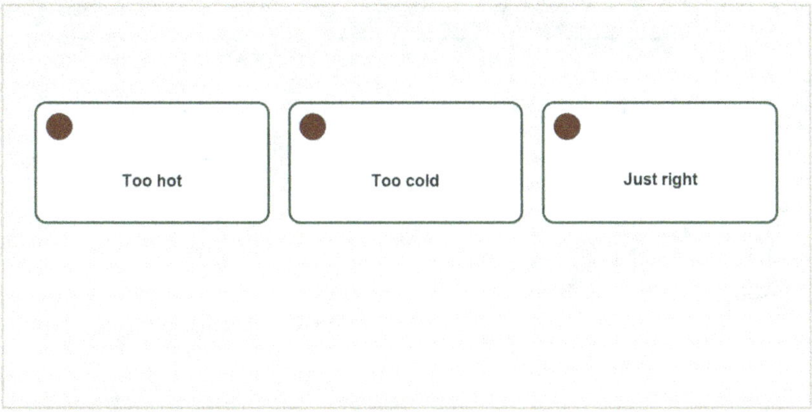

Food and Water

Chicks need chick starter feed, a finely ground ration formulated for their needs at roughly 18 to 20 percent protein, available at all times. Feeding details, including the medicated-versus-unmedicated choice, are covered in the next chapter, but starter is what chicks eat.

Water is critical and carries one specific danger: chicks can drown in even shallow open water, and wet, chilled chicks can die. Use a proper chick waterer with a narrow trough, or add clean pebbles or marbles to a shallow dish so chicks can drink but not fall in or get soaked. Keep water clean and refresh it often, since chicks foul it quickly. When you first introduce chicks to the brooder, gently dip each one's beak in the water so it learns where to drink.

Preventing Chick Drowning in the Waterer

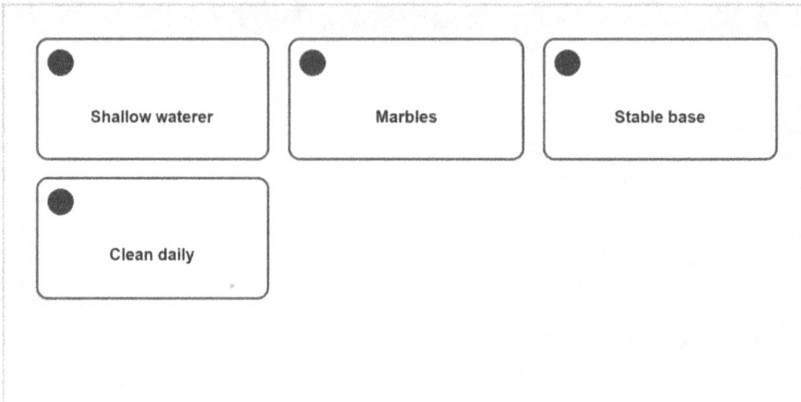

Common Chick Problems

A few issues are common and manageable if caught early. Pasty butt, where droppings cake over and block the vent, is the most frequent; check chicks daily and gently clean any pasted vent with a warm, damp cloth, as a blocked vent can be fatal.

Checking a Chick for Pasty Butt

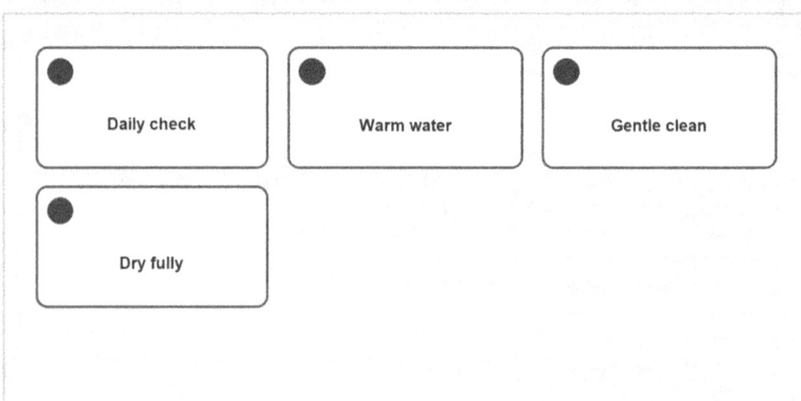

Chilling from a brooder that is too cold or from wet bedding weakens chicks fast, so keep them warm and dry. Pecking among chicks can

start from overcrowding, too much heat, or too little space, and is reduced by giving them more room and proper temperature. Check your chicks closely every day; problems caught early are almost always fixable.

Moving Outside

Chicks are ready to move from the brooder to the coop when they are fully feathered, usually around five to six weeks of age, and when outdoor temperatures are mild enough, ideally similar to the brooder temperature they have reached. Move them to a coop that is ready and predator-proof. If you are adding them to an existing flock, do not simply put them together; integration must be done gradually to manage the pecking order, which is covered in the flock behavior chapter.

Common Mistakes

Using a heat lamp carelessly and risking a fire. Use a brooder plate or secure a lamp thoroughly.

Getting the temperature wrong and not reading chick behavior, leaving chicks too hot or too cold.

Using deep or open water that chicks drown in, or letting them get wet and chilled.

Using cedar bedding or slippery newspaper, harming chicks. Use pine shavings.

Missing pasty butt or other early problems by not checking chicks daily.

Moving chicks outside before they are feathered or into cold weather.

Budget Options

The free or near-free path: Use a repurposed tote or box as a brooder and a brooder plate, which costs little to run, and raise the chicks on inexpensive starter feed and pine shavings.

The starter path: A purpose-made brooder plate, a proper chick waterer and feeder, and a thermometer make raising chicks easy and safe for a modest one-time cost, and the equipment lasts for future batches.

Quick Wins

Set up the brooder and get it to temperature before the chicks arrive.

Use a brooder plate rather than a hot lamp for safety.

Add marbles or pebbles to the chick waterer to prevent drowning.

Plan a daily check for pasty butt and overall health.

Chapter Checklist

[ ] My brooder is set up, draft-free, and at temperature before chicks arrive

[ ] I am using a safe heat source, ideally a brooder plate

[ ] I am following the temperature schedule and reading chick behavior

[ ] My chicks have constant starter feed and safe, shallow water

[ ] I am using pine shavings, not cedar or slippery newspaper

[ ] I check chicks daily for pasty butt and other problems

[ ] I know to move chicks out only when fully feathered

Action Plan

Set up and warm the brooder before the chicks come home, with a safe heat source, pine shavings, starter feed, and a drown-proof waterer. Follow the temperature schedule, reducing heat about five degrees a week while watching the chicks' behavior as your real guide. Check each chick daily for pasty butt and vigor. When they are fully feathered at five to six weeks and the weather is mild, move them to their ready coop. Next, learn exactly how to feed your flock at every age.

Chapter 7: Feeding Your Flock

What You'll Learn

This chapter covers how to feed chickens correctly through their whole life: the right feed for each age, the crucial roles of grit and calcium, what treats and scraps are safe, how much water birds need, and what it all costs. Proper feeding is the foundation of health and steady egg production.

Why It Matters

Feed is your main ongoing cost and the single biggest factor in your flock's health and laying. The right feed at the right age keeps birds healthy and productive, while the wrong feed, especially giving layer feed to young or non-laying birds, can cause real harm. Understanding feed is understanding most of chicken care.

The Right Feed for Each Age

Chickens need different feed as they grow, mainly differing in protein and calcium. Match the feed to the bird's stage.

Chick starter, at roughly 18 to 20 percent protein, feeds chicks from hatch through about six to eight weeks. Grower feed, at roughly 16 to 18 percent protein, feeds growing birds from then until they begin laying, at around eighteen weeks. Layer feed, at roughly 16 percent protein with added calcium, feeds hens from the point of lay onward; the added calcium is essential for forming strong eggshells.

One rule is critical: do not feed layer feed to chicks or to non-laying birds. The high calcium that helps a laying hen can damage the kidneys and development of young or non-laying birds. If you keep a mixed-age flock, a good solution is an all-flock or flock-raiser feed at around 18 to 20 percent protein with lower calcium, fed to everyone, with a separate dish of oyster shell offered so the laying hens can take the extra calcium they need.

Table: What to Feed and When

Stage	Age	Feed	Protein
Chick	0 to 6-8 weeks	Chick starter	About 18 to 20 percent
Grower	6-8 to 18 weeks	Grower feed	About 16 to 18 percent
Layer	18 weeks and laying	Layer feed with calcium	About 16 percent
Mixed-age flock	Any	All-flock plus separate oyster shell	About 18 to 20 percent

Feeding Chickens by Life Stage

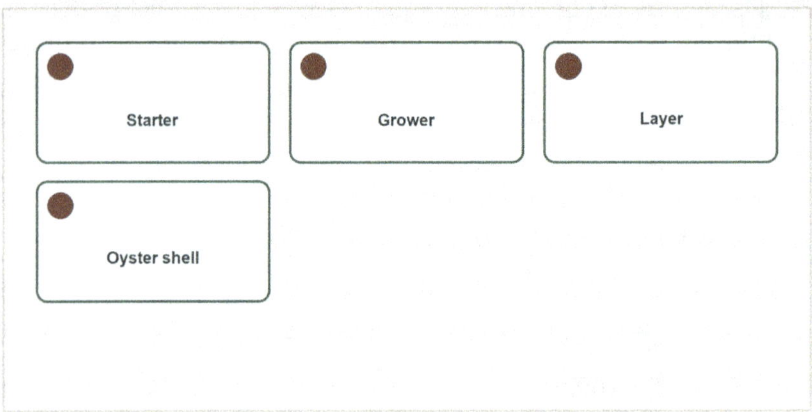

Grit and Calcium: Two Separate Things

Beginners often confuse grit and oyster shell, but they do different jobs and are offered separately, each in its own dish, free choice, so birds take what they need.

Grit is small, hard, insoluble stone that sits in the gizzard and grinds up food, since chickens have no teeth. Birds that eat only commercial crumble or pellet do not strictly need added grit, but any bird that eats grass, scraps, whole grains, or forages needs grit to digest it. Free-ranging birds often pick up enough naturally, but it is cheap insurance to offer it.

Oyster shell is a soluble calcium source for laying hens, helping them form strong eggshells. Offer it free choice in a separate dish, rather than mixing it into feed, so each hen takes the calcium she needs. Thin or soft shells are a classic sign that laying hens need more calcium.

Grit and Oyster Shell - Two Different Jobs

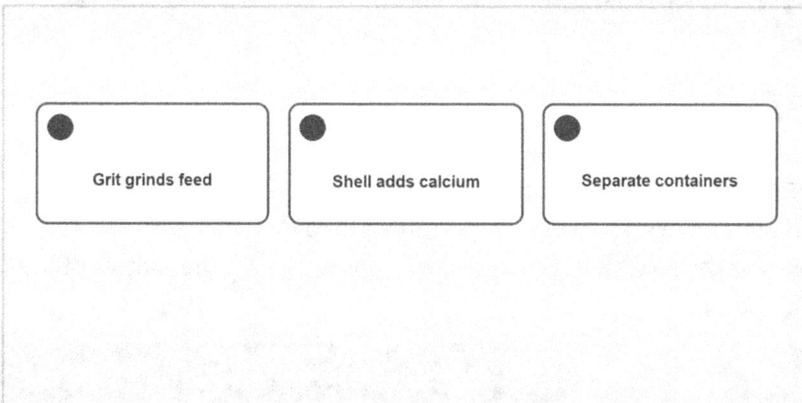

Treats, Scraps, and What to Avoid

Chickens love treats and kitchen scraps, and feeding scraps is one of the joys of keeping them, but treats should make up only a small part of the diet, no more than about a tenth, with balanced feed as the foundation. Too many treats unbalance their nutrition and reduce laying.

Many kitchen and garden scraps are great: vegetable trimmings, fruit, cooked grains, and greens. A few foods should be avoided, including avocado, which is toxic to birds, raw or dried beans, anything moldy or spoiled, very salty or sugary foods, and large amounts of anything that crowds out real feed. Scratch grains, a cracked-corn-and-grain mix, are a treat and a cold-weather warmer, not a complete feed, so offer them sparingly.

Table: Safe Scraps and What to Skip

Category	Examples	Guidance
Good scraps	Vegetable trimmings, fruit, cooked grains, greens	Up to about a tenth of the diet
Treats	Scratch grains, mealworms	Sparingly; not a complete feed
Avoid	Avocado, raw or dried beans, moldy food	Toxic or harmful
Limit	Very salty, sugary, or fatty foods	Unbalances nutrition

Water

Clean, fresh water at all times is as important as feed; a hen that cannot drink stops laying within a day and can sicken quickly. A laying hen drinks roughly a pint of water a day, and considerably more in hot weather, when water becomes a genuine matter of life and death.

Chickens Need Constant Clean Water

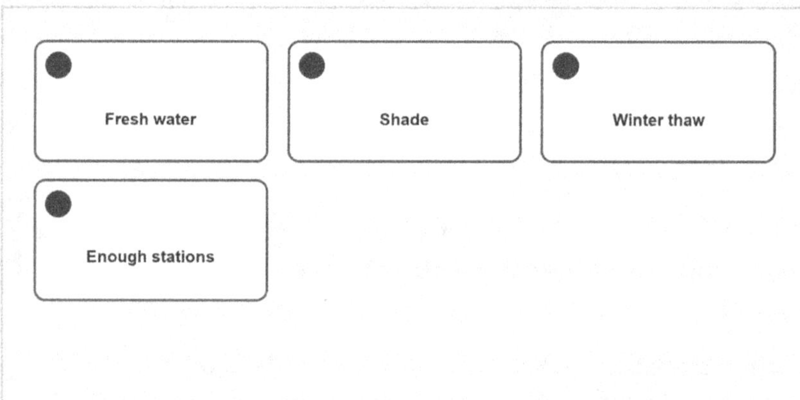

Provide enough waterers that all birds can drink, keep them clean, and in winter prevent freezing with a heated base or by refreshing water regularly. In heat, ensure shade and ample cool water at all times.

What It Costs

Feed is the main recurring cost, and the math is simple and worth knowing. A laying hen eats roughly 1.5 pounds of feed per week, a little under a quarter-pound a day, though this varies with breed, weather, and foraging. For a flock of six hens, that is six times 1.5, which equals 9 pounds per week, or 9 times 52, which is 468 pounds per year.

Feed commonly comes in 50-pound bags. At an example price of 22 dollars per bag, that is 22 divided by 50, or 0.44 dollars per pound, so a year of feed for six hens is 468 times 0.44, which is about 206 dollars, roughly 17 dollars a month, and a single bag lasts that flock about 50 divided by 9, or 5.6 weeks. Plug in your own bag price to update the figure. Birds that forage or eat scraps cost less to feed.

Table: Feed Cost at 1.5 lb per Hen per Week

Hens	Feed per Week	Feed per Year	Cost at 22 per Bag	Cost at 30 per Bag
3 hens	4.5 lb	234 lb	103 dollars	140 dollars
4 hens	6 lb	312 lb	137 dollars	187 dollars
6 hens	9 lb	468 lb	206 dollars	281 dollars
8 hens	12 lb	624 lb	275 dollars	374 dollars

Note: feed per year is hens times 1.5 times 52 weeks; cost is feed per year times the per-pound price, which is 22 divided by 50 (0.44) or 30 divided by 50 (0.60). These figures match those used across this series.

Common Mistakes

Feeding layer feed to chicks or non-laying birds, where the high calcium causes harm. Match feed to stage.

Confusing grit and oyster shell, or mixing calcium into the feed instead of offering it free choice.

Overdoing treats and scraps, unbalancing nutrition and reducing laying.

Feeding toxic foods like avocado or moldy scraps.

Letting water run out or freeze, which stops laying and endangers birds fast.

Storing feed where rodents reach it. Use sealed metal containers.

Budget Options

The free or near-free path: Supplement feed with safe kitchen scraps and garden waste and allow supervised foraging, which can meaningfully cut feed costs, while always keeping balanced feed available as the foundation.

The starter path: Quality age-appropriate feed plus free-choice grit and oyster shell costs little more and keeps birds healthy and laying well, which is the best value over time.

Quick Wins

Match your feed to your birds' current stage right now.

Set out grit and oyster shell in separate free-choice dishes.

Audit your treats so they stay under about a tenth of the diet.

Store feed in a sealed metal container against rodents.

Chapter Checklist

[ ] I feed the correct feed for my birds' age and stage

[ ] I never feed layer feed to chicks or non-laying birds

[ ] I offer grit and oyster shell separately, free choice

[ ] I keep treats and scraps to a small share of the diet and avoid toxic foods

[ ] I provide constant clean water and prevent freezing or shortage

[ ] I store feed securely against rodents

[ ] I understand my flock's feed cost

Action Plan

Match your feed to your flock's stage today, set out separate dishes of grit and oyster shell, and make sure clean water is always available. Keep treats modest and avoid the toxic foods listed. Store feed in sealed metal bins. Work out your monthly feed cost from the table so the expense holds no surprises. With feeding handled, the next chapter turns to the reason most people keep chickens: the eggs.

Chapter 8: Egg Production

What You'll Learn

This chapter explains everything about the eggs: when hens start laying, how many to expect, what makes laying rise and fall, and how to collect, handle, and store eggs safely. You will also learn the truth about a few common questions, including whether you need a rooster and what to do about winter laying.

Why It Matters

Eggs are why most people keep chickens, and understanding production turns confusion into confidence. Knowing what is normal, the slow start, the winter slowdown, the annual molt, means you will not panic when laying changes, and knowing how to handle eggs safely means you get the full, fresh benefit of your flock.

When Hens Start Laying

A pullet typically lays her first egg at around 18 to 22 weeks of age, though it ranges from about 16 to 24 weeks depending on the breed, with lighter laying breeds often starting earlier than heavy ones. The first eggs are usually small, oddly shaped, or have soft shells; this is completely normal and sorts itself out within a few weeks as the hen's system matures. Do not worry about those first quirky pullet eggs.

One fact every beginner should know: hens lay eggs with or without a rooster. A rooster is needed only to fertilize eggs for hatching chicks. For eating eggs, no rooster is required, which is why hens are kept alone in flocks across the country, and why roosters, usually banned in town anyway, are simply unnecessary for an egg flock.

From Pullet to First Egg

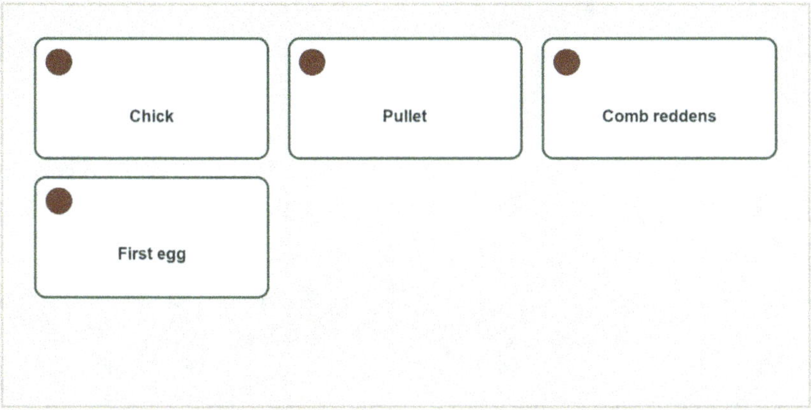

How Many Eggs to Expect

A productive hen of a good laying breed lays roughly four to six eggs a week during her productive season, which adds up to somewhere around 250 or more eggs a year for the best layers. Production is not constant, though; it rises and falls with several natural factors covered below.

A hen lays best in her first two to three years, then production gradually declines, often by something like ten to twenty percent a year, as she ages. She may keep laying for many years, just fewer and fewer eggs. This is why planning for the post-peak years, discussed in Chapter 1, matters.

What Makes Laying Rise and Fall

Several factors drive how many eggs you collect, and understanding them prevents needless worry.

Daylight is the biggest. Laying is triggered by light, and hens lay best with about 14 to 16 hours of daylight. As days shorten in fall and winter, laying naturally slows or stops, then picks up again as days lengthen in spring.

Daylight Drives Laying Through the Year

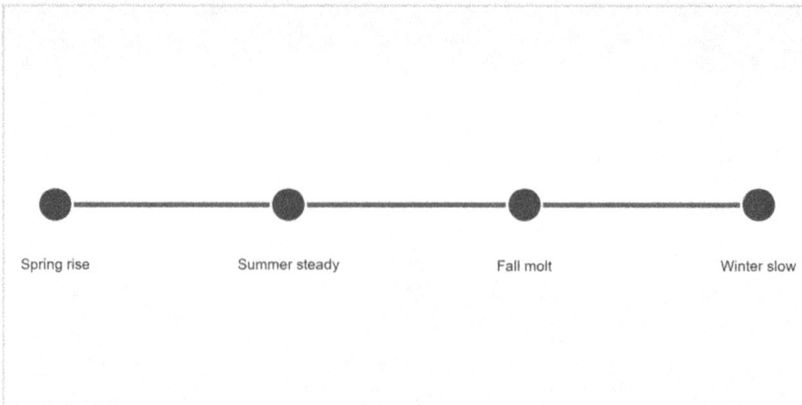

This is normal and seasonal, not a problem.

Age matters, as covered above, with younger hens laying more. The annual molt, described in the seasonal chapter, pauses laying while a hen regrows her feathers. Nutrition affects laying directly; a hen short on feed, calcium, or water lays poorly. Stress from predators, crowding, extreme heat, illness, or disruption can stop laying temporarily. And the breed sets the ceiling; a heavy ornamental breed will never match a production layer no matter how well kept.

Table: Factors That Raise and Lower Egg Production

Factor	Effect	What's Normal
Daylight	Fewer eggs as days shorten	Winter slowdown is natural
Age	Fewer eggs as hens get older	Best in first 2 to 3 years
Molt	Laying pauses during the annual molt	Usually in fall, lasts weeks
Nutrition	Poor feed, calcium, or water lowers laying	Steady supply keeps laying up
Stress	Predators, heat, crowding, illness pause laying	Resolves when stress is removed
Breed	Sets the maximum possible	Choose layers for maximum eggs

The Winter Light Question

Because hens slow down in winter, some keepers add artificial light to the coop to maintain laying through the dark months, extending the day to about 14 to 16 hours. This works, but it involves a genuine trade-off worth understanding. A hen is born with a finite number of potential eggs, and pushing her to lay through winter can use them up faster and may shorten her productive lifespan. Many keepers choose instead to let hens rest naturally in winter, accepting fewer eggs in exchange for healthier, longer-laying birds. If you do add light, add it in the early morning rather than late evening, so birds are not suddenly left in the dark on the roost. There is no single right answer; it is your choice, made with the trade-off in mind.

Collecting and Handling Eggs

Collect eggs daily, ideally more than once a day in very hot or cold weather, to keep them clean, prevent breakage, discourage egg-eating, and stop hens from going broody. Clean nest box bedding means clean eggs and far less washing.

Here is an important and often-misunderstood point about washing and storage. A freshly laid egg has a natural protective coating called the bloom, which seals the shell and keeps the egg fresh. As long as that bloom is intact, meaning the egg is unwashed, the egg stays good for a couple of weeks at room temperature or much longer, often months, in the refrigerator. Once you wash an egg, you remove the bloom, and from then on it must be refrigerated. So the rule is simple: do not wash eggs until you are about to use them, and if you do wash one, refrigerate it. Lightly soiled eggs can be brushed clean dry, saving the bloom.

Collecting and Storing Fresh Eggs

- Collect daily
- Keep clean
- Cool storage
- Use oldest first

The Egg Bloom and Why Washing Matters

- Natural bloom
- Clean nesting
- Wash only if needed
- Refrigerate washed eggs

Checking Freshness

To test an egg's freshness, use the float test. Place the egg in a bowl of water. A very fresh egg sinks and lies flat on the bottom. As an egg ages, it loses moisture and gains air, so an older but still good egg stands upright on the bottom. An egg that floats to the surface is old and should be discarded, since the float means a large air cell and likely spoilage. When in doubt, this simple test settles it.

Table: Judging Egg Freshness

What the Egg Does	Meaning	Use It?
Sinks and lies flat	Very fresh	Yes
Stands upright on the bottom	Older but still good	Yes, use soon
Floats to the surface	Old, likely spoiled	No, discard

Common Egg Problems

A few issues come up. Soft or thin shells usually mean a calcium shortage; offer free-choice oyster shell. A sudden drop or stop in laying is most often seasonal light, molt, or stress, but can signal illness if accompanied by other symptoms. Egg eating, where hens break and eat their own eggs, is a frustrating vice driven by boredom, crowding, or accidental breakage; collect often, keep boxes dark and well-bedded, and ensure good nutrition. Broody hens stop laying to sit on eggs, covered in the behavior chapter.

Common Mistakes

Worrying about small, odd first eggs, which are normal pullet eggs that sort themselves out.

Expecting steady laying through winter and panicking at the natural seasonal slowdown.

Washing eggs immediately and then storing them unrefrigerated, which removes the protective bloom.

Collecting eggs infrequently, leading to dirty, broken, or eaten eggs and broody hens.

Ignoring soft shells instead of providing calcium.

Assuming a rooster is needed for eggs. It is not.

Budget Options

The free or near-free path: Let hens follow their natural seasonal rhythm, resting in winter, which costs nothing and supports long-term health, and rely on clean bedding to keep eggs clean without washing.

The starter path: A few clean, well-placed nest boxes and good free-choice calcium keep eggs clean and shells strong for very little cost, maximizing the eggs you actually get to use.

Quick Wins

Collect eggs daily and leave them unwashed until use.

Offer free-choice oyster shell to keep shells strong.

Use the float test whenever you are unsure of an egg's freshness.

Decide your stance on winter light, knowing the trade-off.

Chapter Checklist

[ ] I understand hens lay without a rooster

[ ] I know to expect first eggs around 18 to 22 weeks, small at first

[ ] I understand laying naturally slows in winter and during molt

[ ] I collect eggs daily and leave them unwashed until use

[ ] I know washed eggs must be refrigerated

[ ] I can use the float test for freshness

[ ] I provide calcium for strong shells

Action Plan

Set a daily egg-collection habit and store your eggs unwashed until you use them, washing only at the point of use. Keep oyster shell available for strong shells, and use the float test whenever freshness is in question. Decide now whether you will add winter light or let your hens rest, with the lifespan trade-off in mind. With production understood, the next chapter keeps your flock healthy enough to keep laying.

Chapter 9: Keeping Your Flock Healthy

What You'll Learn

This chapter covers flock health: preventing illness in the first place, recognizing a sick bird, and handling the common ailments a backyard keeper is likely to meet. You will learn the biosecurity habits that keep disease out, the warning signs to watch for, and when a problem is beyond home care and needs a professional.

Why It Matters

Healthy chickens are mostly the result of good conditions, not constant treatment, so most of flock health is prevention. But illness happens, and chickens instinctively hide weakness, so a sick bird can decline fast before a beginner notices. Knowing the signs and the common problems lets you act early, when action works, and keeps one sick bird from becoming a flock-wide outbreak.

Prevention Comes First

The foundation of flock health is simple and covered throughout this book: a clean, dry, well-ventilated coop; enough space to prevent crowding and stress; clean fresh water and correct feed; and protection from predators and extreme weather. Birds kept this way are remarkably hardy. Most disease in backyard flocks traces back to dampness, crowding, dirty water, or stress, all of which you control.

On top of good conditions, practice basic biosecurity, meaning the habits that keep disease from entering and spreading. Quarantine any new bird away from your flock for about two to four weeks before introducing it, watching for signs of illness, since new birds are the most common way disease arrives. Be cautious about visiting other flocks and then yours without cleaning up, since disease travels on shoes, clothing, and equipment. Keep wild birds and rodents away from feed and water as much as possible. And consider having chicks vaccinated for Marek's disease, a common, serious, and incurable viral illness, which is done at the hatchery on the first day.

Biosecurity Basics - Quarantine, Cleanliness, Pest Control

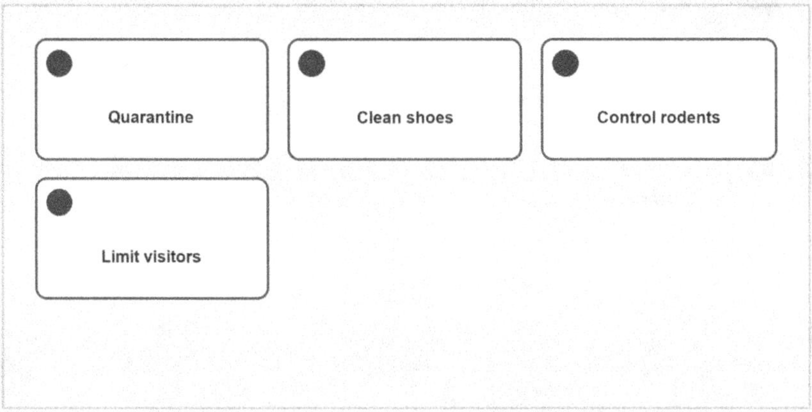

Recognizing a Sick Bird

Because chickens hide illness, you must watch for subtle signs during your daily checks. A sick bird is often lethargic, standing apart from the flock, hunched, or fluffed up. It may eat or drink little, have a pale, shrunken comb, show abnormal droppings such as very watery or bloody ones, breathe with difficulty or make rattling sounds, have a dirty or pasted vent, or simply seem off in a way you notice. Any of these warrants a closer look and, often, separating the bird from the flock. Catching illness early is the single most important factor in treating it successfully.

Daily Health Check of the Flock

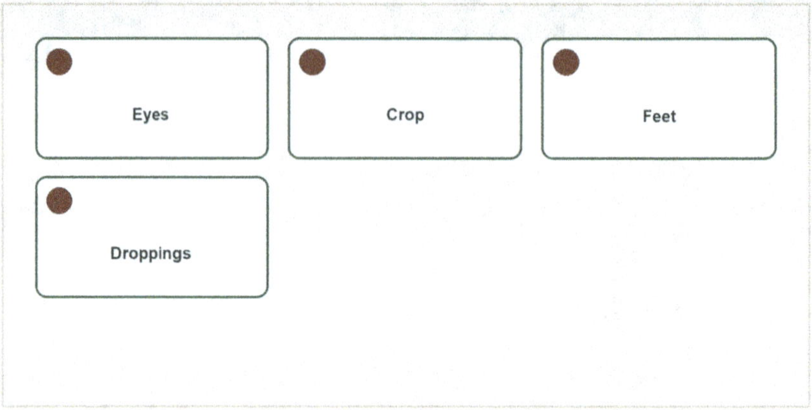

Table: What a Sick Bird Looks Like

Sign	Possible Concern	Action
Lethargy, standing apart, fluffed up	General illness	Isolate and observe closely
Not eating or drinking	Many illnesses, crop issues	Isolate, check crop, monitor
Pale or shrunken comb	Anemia, parasites, illness	Check for mites, assess
Bloody or very watery droppings	Coccidiosis, other illness	Isolate, may need treatment
Labored breathing, rattling, discharge	Respiratory illness	Isolate; can be contagious
Swollen, dirty, or pasted vent	Egg binding, infection, parasites	Examine; egg binding is urgent

Common Ailments

A handful of problems account for most of what backyard keepers encounter.

External parasites, mites and lice, are very common; check around the vent and under feathers for tiny crawling pests or clusters of eggs, and treat with appropriate poultry products while letting birds dust-bathe, which helps them control parasites naturally. Internal parasites, worms, can affect birds and may need deworming if a problem is confirmed. Coccidiosis, an intestinal parasite especially dangerous to chicks, shows as lethargy and bloody droppings; medicated chick starter helps prevent it, and it is treated with a specific medication. Respiratory illnesses cause sneezing, discharge, and rattling breathing, can spread through a flock, and call for isolating the sick bird. Bumblefoot is a foot infection causing swelling and limping, often from a cut or rough landing. Egg binding, when an egg gets stuck inside a hen, is an urgent emergency shown by straining, distress, and a hunched posture. Frostbite can affect combs and wattles in cold, damp conditions, which good ventilation and dryness help prevent.

You do not need to memorize every disease. Watch for the warning signs, isolate any sick bird, address parasites and conditions, and seek help for anything serious or unfamiliar.

A Basic First Aid Kit

Keep a few basics on hand so you are not scrambling when something happens: a poultry-safe wound spray or antiseptic, clean gauze and self-adhesive bandage wrap, an appropriate poultry parasite treatment, electrolyte and vitamin supplements for water to support stressed birds, gloves, and a dedicated container to isolate a sick or injured bird. Simple, prepared, and ready.

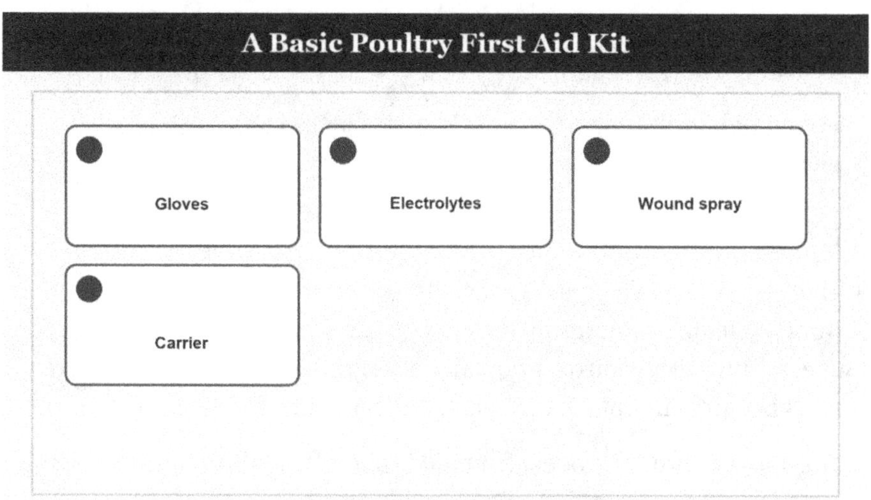

When to Get Help

Home care handles minor issues, but know your limits. For serious illness, an injury beyond a minor wound, anything spreading through the flock, or any condition you cannot identify, consult a veterinarian who treats poultry or your local agricultural extension service, which often helps identify problems and recommend region-appropriate solutions. One important safety note: if you ever treat a bird with medication, observe any required withdrawal period before eating that bird's eggs or meat, as stated on the product, so no residue reaches your food.

Common Mistakes

Treating health as something to react to rather than prevent. Conditions and biosecurity prevent most illness.

Skipping quarantine for new birds, the most common way disease enters a flock.

Missing early signs because of infrequent observation. Check the flock daily.

Leaving a sick bird in the flock, letting contagious illness spread.

Ignoring withdrawal periods after medicating, risking residue in eggs or meat.

Trying to handle a serious or unknown illness alone instead of seeking help.

Budget Options

The free or near-free path: Prevention is free and does almost all the work: a clean, dry, uncrowded coop, fresh water, correct feed, quarantine of new birds, and daily observation. A dust-bathing area costs nothing and helps birds manage parasites themselves.

The starter path: A modest first aid kit and the willingness to consult a vet or extension service when needed cover the situations prevention cannot, protecting both individual birds and the whole flock.

Quick Wins

Start a daily habit of observing the flock for the warning signs.

Set up a quarantine plan for any new birds.

Provide a dry dust-bathing area to help birds control parasites.

Assemble a basic poultry first aid kit.

Chapter Checklist

[ ] I rely on clean, dry, uncrowded conditions to prevent most illness

[ ] I quarantine new birds for two to four weeks

[ ] I observe the flock daily for warning signs

[ ] I isolate any sick bird promptly

[ ] I have a basic first aid kit

[ ] I know when to consult a vet or extension service

[ ] I will observe medication withdrawal periods

Action Plan

Lock in prevention first: keep the coop clean, dry, ventilated, and uncrowded, and commit to observing your flock closely every day. Set up a quarantine space for any future new birds and assemble a simple first aid kit. Learn the warning signs so you can isolate and act early, and identify a poultry vet or extension contact now, before you need one. With health covered, the next chapter helps your flock thrive through every season.

Chapter 10: Seasonal Care

What You'll Learn

This chapter covers caring for your flock through the year: keeping birds safe and laying in winter cold, protecting them from dangerous summer heat, understanding the annual molt, and the rhythm of spring. Chickens are seasonal creatures, and matching your care to the season keeps them healthy and productive.

Why It Matters

The biggest seasonal risks are the ones beginners get backwards. Many worry about cold and overheat their coops, creating fire and moisture hazards, while underestimating heat, which is far more deadly to chickens. Understanding the real seasonal dangers, and the genuinely helpful responses, protects your flock when conditions turn extreme.

Winter Care

Chickens handle cold far better than most beginners expect. They wear a down coat, huddle together, and tuck into their feathers, and most cold-hardy breeds tolerate freezing temperatures well as long as they stay dry and out of drafts. The key winter truths run against intuition.

Do not add a heat lamp to the coop. This is important. Heaters are a serious fire risk in a dusty, feathered coop, and worse, they prevent birds from acclimating to the cold, so a power failure or a bird locked out can then be deadly. A dry, draft-free, well-ventilated coop full of birds stays workable through hard winters without added heat. Keep ventilation going even in winter, because trapped moisture, not cold, causes frostbite and respiratory illness. Prevent the water from freezing with a heated waterer base or by swapping in fresh water regularly, since birds must drink. A deeper layer of bedding adds insulation and, with the deep litter method, generates a little gentle warmth as it composts. Offer a handful of scratch grains in the evening, since digesting them overnight produces body heat. And in extreme cold, check combs and wattles for frostbite, which good ventilation and dryness help prevent.

Winter Coop Care - Dry, Ventilated, No Heat Lamp

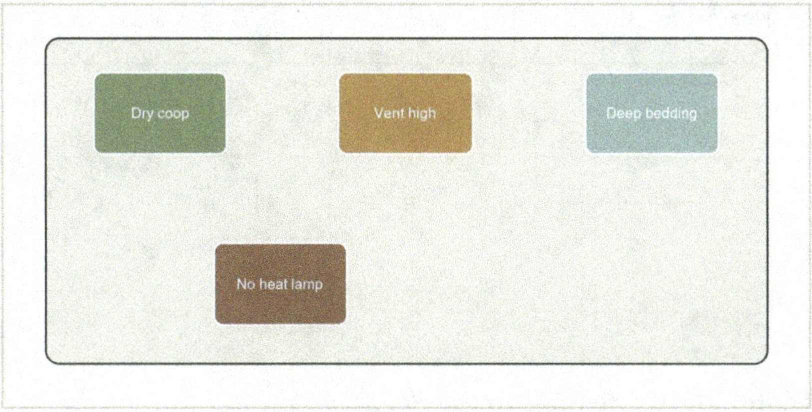

Summer and Heat

Heat is more dangerous to chickens than cold, and heat stress can kill quickly, so summer care is genuinely urgent in hot climates. Chickens cannot sweat; they cool themselves by panting and by holding their wings away from their bodies, and when those are not enough, they fail fast.

Provide constant shade so birds can always escape the sun. Supply ample cool, fresh water, more than you think, and consider several waterers in hot spells. Ensure good airflow through the coop and run. On extreme days, offer cooling help such as frozen treats, cool produce, or a shallow pan of water for their feet, and a dust-bathing area in the shade. Watch closely for heat stress, shown by heavy panting, wings held out, lethargy, pale combs, and reduced eating and laying, and act immediately to cool any affected bird. Avoid handling or disturbing birds during the heat of the day. In hot regions, summer water and shade are matters of life and death.

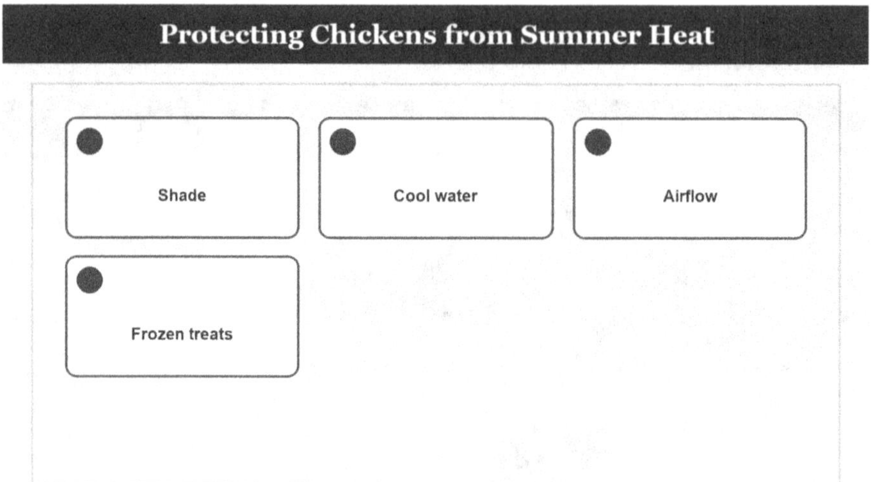

Table: Matching Care to the Season

Season	Main Risk	Key Actions
Winter	Damp and drafts, not cold itself	Keep dry, ventilate, prevent frozen water, no heat lamp
Summer	Heat stress, which is deadly	Shade, ample cool water, airflow, cooling help
Fall	The molt, lower laying	Higher protein, reduce stress, expect fewer eggs
Spring	Broodiness, resumed laying	Watch for broody hens, enjoy the egg surge

The Molt

Once a year, usually in fall, chickens go through a molt, shedding and regrowing their feathers. A molting bird can look ragged, even alarmingly patchy, and will usually stop laying while the molt runs its course over several weeks to a couple of months, because regrowing feathers takes the same protein that eggs require. This is normal and healthy. Support molting birds with a higher-protein feed to help feather regrowth, avoid handling them since new pin feathers are tender and sensitive, and minimize stress. Laying resumes once the molt is complete and, often, once the days lengthen again.

A Hen During the Annual Molt

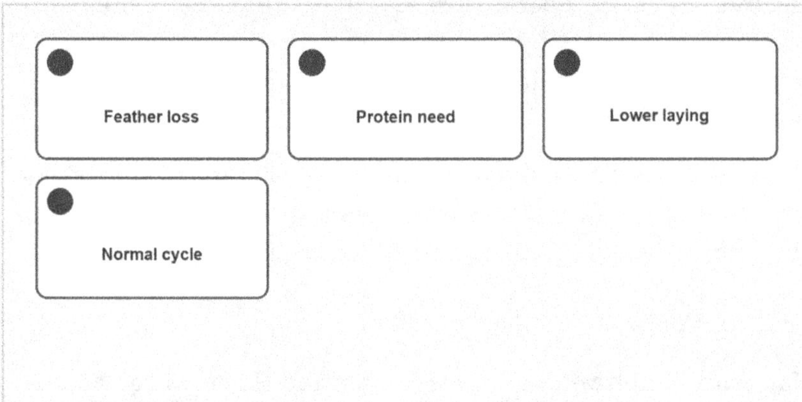

Spring

Spring is the flock's busy, productive season. Lengthening days restart and boost laying, often producing a happy surge of eggs. It is also when broodiness appears, as some hens feel the urge to sit on eggs, covered in the next chapter, and the classic season for adding chicks to grow the flock. Use the mild weather for a thorough coop clean-out after winter, and enjoy the return of abundant eggs.

Common Mistakes

Adding a heat lamp in winter, creating a fire risk and preventing birds from acclimating to cold.

Sealing the coop tight in winter and trapping the moisture that actually causes frostbite.

Underestimating summer heat, the deadliest seasonal threat, by failing to provide shade and ample water.

Panicking at the molt and assuming a ragged, non-laying bird is sick.

Letting winter water freeze, leaving birds unable to drink.

Budget Options

The free or near-free path: Most seasonal care is free: ventilation and dry deep bedding in winter, shade and extra water containers in summer, and patience through the molt. These cost little or nothing and address the real risks.

The starter path: A heated waterer base for winter and extra shade structures or waterers for summer are inexpensive and remove the two genuine seasonal dangers, frozen water and heat stress.

Quick Wins

Make sure winter ventilation stays open and skip the heat lamp.

Set up reliable shade and extra water before the first heat wave.

Stock a higher-protein feed for molting season.

Plan a thorough coop clean-out for spring.

Chapter Checklist

[ ] My winter coop is dry, ventilated, and unheated

[ ] I prevent the water from freezing in winter

[ ] I provide ample shade and cool water against summer heat

[ ] I recognize heat stress and act on it immediately

[ ] I understand the molt and support birds through it

[ ] I use spring for a thorough coop clean-out

Action Plan

Prepare for the season ahead now rather than during a crisis. For winter, ensure ventilation and a way to keep water unfrozen, and resist the urge to add heat. For summer, set up dependable shade and extra water before the heat arrives, and learn the signs of heat stress. Expect and support the fall molt, and use spring's mild days to deep-clean the coop. With seasonal care handled, the next chapter helps you manage the flock's social life and behavior.

Chapter 11: Flock Behavior and Management

What You'll Learn

This chapter explains how chickens behave as a flock and how to manage them: the pecking order, safely introducing new birds, handling broody hens, the question of roosters, and solving problem behaviors like feather-picking and egg-eating. Understanding flock behavior turns confusing squabbles into something you can read and manage.

Why It Matters

Chickens are social animals with their own rules, and most of what beginners find alarming, the pecking, the squabbling, a hen suddenly refusing to leave the nest, is normal flock behavior. Knowing what is normal and what needs intervention keeps your flock peaceful and lets you add birds, manage broodiness, and stop harmful habits before they spread.

The Pecking Order

Chickens live by a social hierarchy literally called the pecking order, where each bird knows its rank, with higher-ranking birds getting first access to food, water, and the best roost spots. The flock establishes and maintains this order through pecking and posturing, which looks rougher than it is. A certain amount of chasing, squabbling, and minor pecking is completely normal and how the flock keeps the peace.

Intervention is needed only when it turns harmful, especially if a bird draws blood, because the sight of blood triggers others to pile on, and an injured bird can be pecked relentlessly, even to death. If a bird is bleeding or being singled out and bullied badly, separate it to heal and address the underlying cause, usually crowding, boredom, or a new-bird introduction gone wrong.

The Pecking Order in a Flock

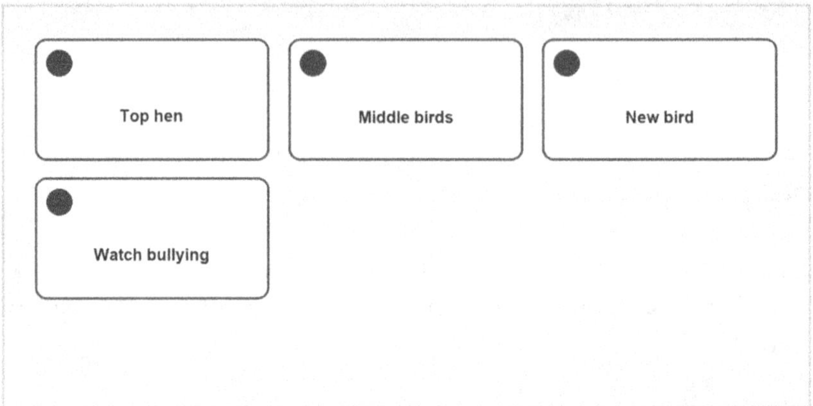

Introducing New Birds

Adding birds to an established flock must be done carefully, because the existing flock sees newcomers as intruders and the pecking order has to be renegotiated. Never simply drop new birds in with the others; it can lead to vicious bullying.

Follow two stages. First, quarantine new birds separately for two to four weeks for health, as covered in Chapter 9. Then integrate gradually: house the new birds where the flock can see but not touch them, such as in an adjacent pen or a wire partition, for about a week so they get familiar without fighting. When you do mix them, do it with plenty of space and distractions, and many keepers add new birds at night onto the roost, so the flock wakes up to the newcomers already present, which can ease the transition. Expect some squabbling as the order resets; intervene only if it turns bloody. Patience here prevents real harm.

Introducing New Birds Gradually

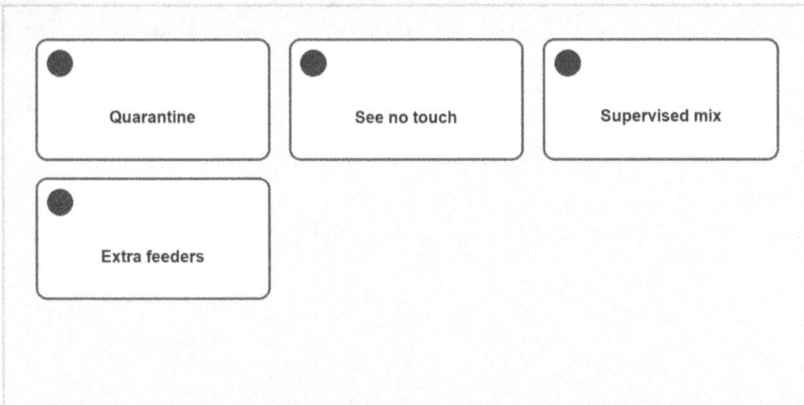

Broody Hens

Sometimes a hen goes broody, overcome by the instinct to sit on eggs and hatch them. A broody hen stays planted in the nest box, fluffs up and may growl or peck when disturbed, plucks her breast feathers, and stops laying. Some breeds, like Orpingtons and Silkies, go broody readily; production hybrids rarely do.

You have two choices. If you want chicks and have fertile eggs, meaning you have a rooster, a broody hen is a gift; she will happily hatch and raise a clutch with no incubator needed. If you do not want chicks, which is the usual case for an egg flock with no rooster, you should break the broodiness, because a broody hen does not lay and can lose condition sitting for weeks. Break it by repeatedly removing her from the nest, blocking access to the nest box, and cooling her underside, often by placing her in a separate wire-floored cage with food and water but no bedding to nest in for a few days, which interrupts the hormonal cycle. The sooner you act, the easier it is.

Recognizing and Managing a Broody Hen

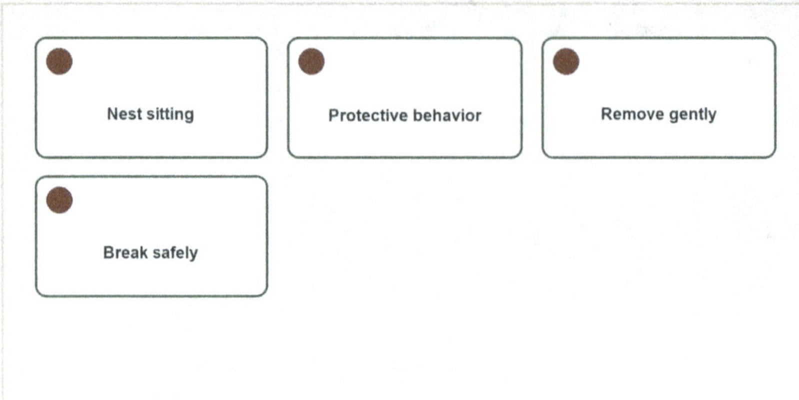

The Question of Roosters

Roosters are optional and, for most backyard keepers, unnecessary and often illegal. On the plus side, a good rooster watches for predators and alerts the flock, helps keep order, and is required if you want fertile eggs to hatch chicks. On the minus side, roosters crow loudly and often, including well before dawn, some become aggressive toward people, and they are banned in most residential areas. Remember the central fact: you do not need a rooster for eggs. Keep one only if you want chicks, your rules allow it, and you are prepared for the noise and the occasional attitude.

Solving Problem Behaviors

A few behaviors cause real trouble and usually trace back to crowding, boredom, or a nutritional gap.

Feather-picking and cannibalism, where birds pull each other's feathers or worse, most often come from too little space, boredom, too much light, or a protein deficiency. Address it by giving more room, adding enrichment, ensuring adequate protein, reducing excessive light, and removing any injured bird until healed, since blood draws attackers. Egg-eating, where hens break and eat their own eggs, starts from boredom, crowding, accidental breakage, or calcium shortage; combat it by collecting eggs often, keeping nest boxes dark and well-bedded, providing calcium, and removing any habitual offender's opportunity. Bullying beyond normal pecking signals crowding or a mismatched flock and is eased with more space and resources, such as multiple feeding and watering stations so lower-ranked birds can eat in peace.

The common thread is that most problem behaviors are prevented by enough space, something to do, and good nutrition. Bored, crowded chickens invent bad habits.

Table: Common Vices and How to Fix Them

Behavior	Usual Cause	Solution
Feather-picking	Crowding, boredom, low protein, too much light	More space, enrichment, protein, less light
Egg-eating	Boredom, breakage, low calcium	Collect often, dark boxes, calcium
Bullying beyond normal	Crowding, mismatched flock	More space, multiple food and water stations
Excessive broodiness	Breed instinct	Break broodiness or let her hatch chicks

Enrichment and Taming

A little enrichment prevents boredom and the vices it causes: things to peck and scratch at, a hanging vegetable, a dust-bathing area, perches, and room to forage. Dust bathing, where chickens roll in dry loose dirt, is a natural and essential behavior that helps them control parasites, so always provide a dry spot for it. To tame and bond with your flock, spend calm time around them, move gently, and offer occasional treats from your hand; chickens raised with friendly daily contact become remarkably tame and enjoyable.

Common Mistakes

Panicking at normal pecking-order behavior, which is how the flock keeps peace.

Dropping new birds straight into the flock instead of integrating them gradually.

Leaving a bleeding bird in the flock, where others will attack it.

Letting a broody hen sit for weeks when you do not want chicks, losing eggs and her condition.

Keeping a rooster without need, dealing with noise, aggression, and possible illegality.

Ignoring the crowding, boredom, or nutrition behind problem behaviors and treating only the symptom.

Budget Options

The free or near-free path: Space, enrichment, and good management cost little or nothing and prevent most behavior problems. A dust-bathing area, scattered scratch to forage for, and a simple wire partition for introductions are nearly free.

The starter path: Multiple feeders and waterers and a separate small pen for quarantine, integration, or breaking broodiness make flock management easy and are inexpensive, lasting tools.

Quick Wins

Provide a dry dust-bathing area today; it is natural and prevents parasites.

Set up multiple food and water stations so lower-ranked birds can eat in peace.

Plan a see-but-not-touch method for any future bird introductions.

Spend a few calm minutes with your flock to begin taming them.

Chapter Checklist

[ ] I understand normal pecking-order behavior and when to intervene

[ ] I introduce new birds gradually, never by dropping them in

[ ] I know how to identify and handle a broody hen

[ ] I understand roosters are optional and unnecessary for eggs

[ ] I prevent problem behaviors with space, enrichment, and nutrition

[ ] I provide a dust-bathing area and multiple feeding stations

Action Plan

Set your flock up for social harmony: provide a dust-bathing area, multiple feeding and watering stations, and some enrichment to prevent boredom. Learn the see-but-not-touch method so you can add birds safely later, and know the signs of a broody hen and how to respond. Spend calm time with your birds to tame them. With behavior understood, the final chapter pulls everything together into a first-year roadmap.

Chapter 12: The Chicken Year and First-Year Roadmap

What You'll Learn

This final chapter ties the whole book together into a month-by-month roadmap for your first year, from bringing home chicks to collecting eggs through your first winter. You will see what to expect at each stage, the seasonal rhythm of flock-keeping, and how to think about growing your flock in the years ahead.

Why It Matters

You now understand every part of keeping chickens. This roadmap sequences it so you do the right thing at the right time and know what is coming, which removes the uncertainty that makes beginners anxious. With a clear picture of the first year, you can start with confidence.

The First Year, Month by Month

The roadmap below assumes the most common start: day-old chicks brought home in spring. Adjust the timing if you start with grown hens, which skip straight to laying, or begin in another season.

Table: From Chicks to Eggs, Month by Month

Stage	Timing	What's Happening
Bring home chicks	Early spring	Brooder set up; chicks need careful heat and daily care
Brooding	Weeks 1 to 5	Reduce heat weekly; chicks feather out; prepare the coop
Move to coop	Around 6 weeks	Feathered pullets go to a ready, predator-proof coop
Growing	About 6 to 16 weeks	Grower feed; predator vigilance; summer heat care
Point of lay nears	About 16 to 18 weeks	Switch to layer feed; nest boxes ready
First eggs	About 18 to 22 weeks	Small pullet eggs at first, then steady laying
First fall and winter	Months later	Spring pullets often lay through their first winter
Second spring	About one year	Full production; possible broodiness; consider expanding

First-Year Chicken Roadmap Timeline

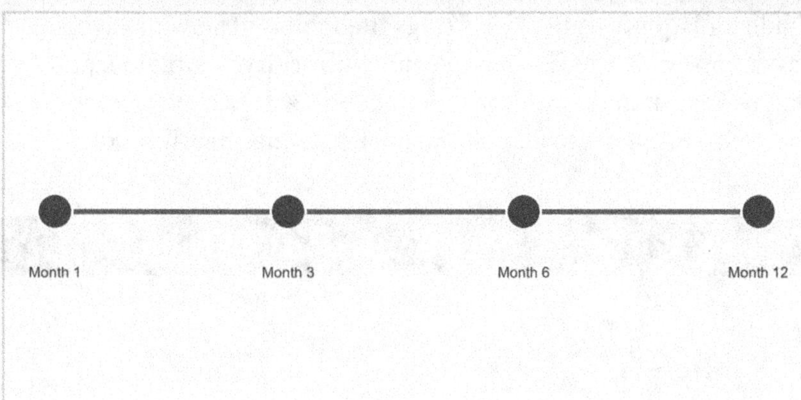

What to Expect Along the Way

A few things surprise first-year keepers, so expect them. The brooder weeks feel intense but pass quickly. Pullets grow faster than you think and will be ready for the coop around six weeks. The wait for the first egg can feel long, around five to six months from chicks, so be patient. First eggs are small and sometimes odd, which is normal. And here is a pleasant surprise unique to the first year: pullets that begin laying in late summer or fall usually lay right through their first winter, because they do not go through their first molt until they are around eighteen months old. So your first winter often brings eggs even as older hens elsewhere slow down. The following fall is when you will see your birds' first real molt and winter slowdown.

The Seasonal Rhythm

After the first year, flock-keeping settles into a yearly rhythm worth internalizing. Spring brings rising laying, broodiness, and the season for adding new chicks. Summer demands vigilance against heat, with shade and ample water. Fall brings the molt and slowing laying as days shorten. Winter is about keeping birds dry, ventilated, and watered through the cold, without added heat. Each year you repeat the cycle with more skill, and the flock becomes nearly second nature.

The Chicken Year Seasonal Rhythm

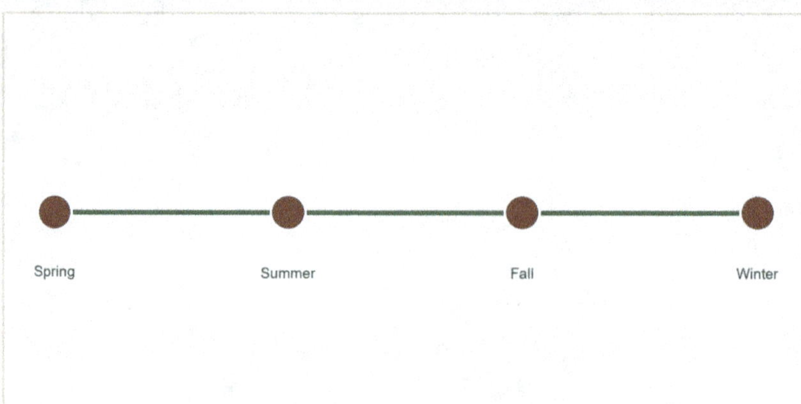

Growing Your Flock

Once your first year is behind you and the routine feels easy, you may want to expand. You can add more hens, integrating them carefully as covered in Chapter 11, to increase your egg supply. You might branch into other poultry such as ducks or quail, which is the territory of Book 7, Small Livestock for Beginners. Some keepers eventually raise meat birds or hatch their own chicks with a rooster and a broody hen or incubator, both more advanced steps to grow into. Whatever direction you choose, the foundation you build this first year, healthy birds, a safe coop, good feeding, and confident management, supports all of it.

Remember, too, how chickens fit the wider homestead: their manure enriches the garden beds from Book 2, they happily turn kitchen and garden scraps into eggs, and their bedding feeds the compost pile of Book 5. The flock is not a standalone project but a hub that connects to everything else you grow.

Common Mistakes

Expecting eggs too soon and growing discouraged during the months-long wait from chicks.

Forgetting to switch from grower to layer feed as pullets approach laying age.

Not having the coop ready by the time chicks are feathered at around six weeks.

Assuming first-winter pullets will slow down like older hens, when they often lay right through.

Rushing to expand the flock before the first-year routine is solid.

Budget Options

The free or near-free path: The first-year roadmap costs nothing to follow; it is simply doing the right task at the right time. Starting with a small flock of chicks in spring is the most economical path through the whole year.

The starter path: Investing in the things that make each stage easy, a safe brooder, a solid coop, and reliable feeders and waterers, pays off across the first year and every year after, as covered in their chapters.

Quick Wins

Map your own first year onto the roadmap based on when you start.

Mark the feed-switch point as pullets near laying age.

Make sure the coop is ready before chicks reach six weeks.

Plan to enjoy first-winter eggs from spring pullets.

Chapter Checklist

[ ] I have mapped the first-year roadmap to my own start

[ ] I know the brooder stage is intense but brief

[ ] I will have the coop ready by the time chicks are feathered

[ ] I will switch to layer feed as pullets near laying

[ ] I expect first eggs around five to six months in

[ ] I understand spring pullets often lay through their first winter

[ ] I will wait until the routine is solid before expanding

Action Plan

Map the roadmap onto your own calendar based on when and how you are starting, marking the key transitions: moving feathered chicks to the coop around six weeks, switching to layer feed near eighteen weeks, and expecting first eggs around five to six months. Follow the seasonal rhythm thereafter, and resist expanding until your routine feels easy. You now have everything you need to raise a healthy, productive flock from day-old chick to fresh eggs and beyond.

Conclusion: Your Thriving Flock

You started this book with an empty yard and questions. You now know how to handle the legal groundwork, choose the right breeds, build a safe coop and predator-proof run, bring home and raise healthy birds, feed them correctly at every age, collect and store fresh eggs, keep the flock healthy, and manage them through every season and situation. That is a complete chicken-keeping education, and it is yours.

What to Do First

Do not let this knowledge wait. Take the smallest concrete step today: confirm your legal hen limit, choose your breeds, or price out a coop. The first real action turns reading into a flock. And remember that the first egg from your own hen, which arrives a few patient months after you begin, is one of the small, genuine joys of self-sufficiency.

Where This Fits in Your Homestead

A backyard flock is one of the most rewarding pieces of a homestead, and it connects to everything else in this series. The coop and run you need are built with checked plans in Book 6, Homestead Projects. The eggs you cannot use fresh can be preserved using Book 4, Food Preservation. Your hens' manure and bedding feed the compost in Book 5, Composting Made Simple, which in turn feeds the raised beds of Book 2, Raised Bed Gardening, whose scraps come back to the hens. When you are ready for other animals, Book 7, Small Livestock for Beginners, covers ducks, quail, and rabbits. The flock sits at the heart of a working homestead.

A Final Word

Keeping chickens is one of the most accessible and satisfying steps toward self-sufficiency anyone can take. You do not need a farm or any experience, just a safe coop, good food and water, protection from predators, and a few minutes of care each day. The birds do the rest, and they reward you every morning. You have the knowledge. Go get your flock.

About the Author

Hudson Briggs is a lifelong DIY enthusiast, gardener, and homesteading educator dedicated to helping everyday people become more self-sufficient. Through practical guidance, realistic budgets, and beginner-friendly instruction, he helps readers build skills that lead to greater independence, confidence, and resilience.

The Homestead Starter Series

Book 1 — The Beginner's Guide to Homesteading

Book 2 — Raised Bed Gardening

Book 3 — Backyard Chickens

Book 4 — Food Preservation

Book 5 — Composting Made Simple

Book 6 — Homestead Projects

Book 7 — Small Livestock for Beginners

Book 8 — Homestead Budgeting

Disclaimer

Disclaimer: The information in this book is provided for general educational purposes and reflects widely accepted poultry-keeping practices and realistic averages at the time of writing. Local laws, climate, breeds, prices, and conditions vary and change over time, so always confirm your own local regulations and adjust for your situation. For any sick or injured bird beyond basic care, consult a veterinarian or your local agricultural extension service, and always observe medication withdrawal periods before consuming eggs or meat. This book is not a substitute for professional veterinary advice.

www.ingramcontent.com/pod-product-compliance
Lightning Source LLC
LaVergne TN
LVHW022311200926
848481LV00004B/1277
9798995255840